AF294391

Springer Theses

Recognizing Outstanding Ph.D. Research

For further volumes:
http://www.springer.com/series/8790

Aims and Scope

The series "Springer Theses" brings together a selection of the very best Ph.D. theses from around the world and across the physical sciences. Nominated and endorsed by two recognized specialists, each published volume has been selected for its scientific excellence and the high impact of its contents for the pertinent field of research. For greater accessibility to non-specialists, the published versions include an extended introduction, as well as a foreword by the student's supervisor explaining the special relevance of the work for the field. As a whole, the series will provide a valuable resource both for newcomers to the research fields described, and for other scientists seeking detailed background information on special questions. Finally, it provides an accredited documentation of the valuable contributions made by today's younger generation of scientists.

Theses are accepted into the series by invited nomination only and must fulfill all of the following criteria

- They must be written in good English.
- The topic should fall within the confines of Chemistry, Physics, Earth Sciences, Engineering and related interdisciplinary fields such as Materials, Nanoscience, Chemical Engineering, Complex Systems and Biophysics.
- The work reported in the thesis must represent a significant scientific advance.
- If the thesis includes previously published material, permission to reproduce this must be gained from the respective copyright holder.
- They must have been examined and passed during the 12 months prior to nomination.
- Each thesis should include a foreword by the supervisor outlining the significance of its content.
- The theses should have a clearly defined structure including an introduction accessible to scientists not expert in that particular field.

Kipp van Schooten

Optically Active Charge Traps and Chemical Defects in Semiconducting Nanocrystals Probed by Pulsed Optically Detected Magnetic Resonance

Doctoral Thesis accepted by the
University of Utah, USA

 Springer

Kipp van Schooten
Department of Physics & Astronomy
University of Utah
Salt Lake City
Utah, USA

Supervisor
Christoph Boehme
Department of Physics & Astronomy
University of Utah
Salt Lake City
Utah, USA

ISSN 2190-5053 ISSN 2190-5061 (electronic)
ISBN 978-3-319-03328-0 ISBN 978-3-319-00590-4 (eBook)
DOI 10.1007/978-3-319-00590-4
Springer Cham Heidelberg New York Dordrecht London

© Springer International Publishing Switzerland 2013
Softcover re-print of the Hardcover 1st edition 2013
This work is subject to copyright. All rights are reserved by the Publisher, whether the whole or part of the material is concerned, specifically the rights of translation, reprinting, reuse of illustrations, recitation, broadcasting, reproduction on microfilms or in any other physical way, and transmission or information storage and retrieval, electronic adaptation, computer software, or by similar or dissimilar methodology now known or hereafter developed. Exempted from this legal reservation are brief excerpts in connection with reviews or scholarly analysis or material supplied specifically for the purpose of being entered and executed on a computer system, for exclusive use by the purchaser of the work. Duplication of this publication or parts thereof is permitted only under the provisions of the Copyright Law of the Publisher's location, in its current version, and permission for use must always be obtained from Springer. Permissions for use may be obtained through RightsLink at the Copyright Clearance Center. Violations are liable to prosecution under the respective Copyright Law.
The use of general descriptive names, registered names, trademarks, service marks, etc. in this publication does not imply, even in the absence of a specific statement, that such names are exempt from the relevant protective laws and regulations and therefore free for general use.
While the advice and information in this book are believed to be true and accurate at the date of publication, neither the authors nor the editors nor the publisher can accept any legal responsibility for any errors or omissions that may be made. The publisher makes no warranty, express or implied, with respect to the material contained herein.

Printed on acid-free paper

Springer is part of Springer Science+Business Media (www.springer.com)

*To my wife, Teatha, who is a whip and a
wonder.*

Supervisor's Foreword

An important focus of contemporary nanoscience is the investigation and development of semiconducting colloidal nanocrystals for optoelectronic device applications. Being highly facile in their synthesis, a wide range of sizes, morphologies, materials, interactions, and effects can be engineered by synthetic chemists. The solution-processability of these systems also allow the use of long established industrial fabrication techniques such as reel-to-reel processing or even simple inkjet printing, offering the prospect of extremely low-cost device manufacturing. Aside from these anticipated technologies, this material class also opens up a "playground" for generating and observing previously unseen quantum effects of reduced dimensional systems.

The surface-to-volume ratio of nanocrystals is very large and therefore, unsatisfied surface states are able to significantly influence a nanocrystals density of electronic states. Passivation methods for these states exist, yet these are not fully effective, often due to a significant change in the surface stoichiometry caused by complex atomic reorganization during passivation. Surface states are charge traps and their detrimental effect on optical observables is readily seen, for instance, in single particle photoluminescence (PL) blinking. One step towards a control of such detrimental behavior is the development of an understanding of the physical and chemical nature of these states. Unfortunately, there is only a limited set of available experimental methods which allow the observation of these trap states. In absence of structural information though, systematically engineering a robust nanocrystal surface passivation becomes problematic.

Kipp van Schooten's thesis focuses on the application of pulsed optically detected magnetic resonance (pODMR) for the direct access to the physical and chemical nature of optically active charges under trapping conditions. By application of pODMR, Kipp has provided a wealth of knowledge about the identity and the local microscopic environment of these states, while also clarifying their role for excitonic processes, which are responsible for charge to light conversion in these materials. Kipp's work has been able to show for the first time that both electrons and holes are able to be trapped within the same nanoparticle at the same time, and that they form weakly spin exchange coupled, excitonic precursor pairs. Further, by taking advantage of the extraordinary long spin coherence lifetime (T_2) for these localized charge traps, Kipp accomplished the demonstration of new effects, such as the coherent spin control of the light-harvesting yield and remote readout of spin information in CdSe/CdS heterostructure nanocrystals.

Kipp has demonstrated the surprising and counterintuitive phenomena of very long spin coherence lifetimes, of up to $T_2 \approx 1.6\,\mu s$, for trap and defect states in these high atomic number materials which intrinsically possess large spin-orbital coupling. This opens up the possibility of highly precise chemical fingerprinting through use of advanced spin resonance techniques, such as optically detected electron spin echo envelop modulation (ESEEM).

Kipp van Schooten's dissertation lays the groundwork for further use of pODMR, and more powerful pulsed magnetic resonance probes for the investigations of electronic states that fundamentally limit the practical usability of colloidal nanocrystal optoelectronics devices. Furthermore, by gaining access to these optically active electronic states, novel methods of coherent quantum control may be tested and further developed and this could lead to technological developments far beyond the scope of lighting and light harvesting technologies.

Salt Lake City, UT, USA Christoph Boehme

Acknowledgments

I had been leaving this acknowledgments section to the very last of the work needed to be done while finishing this dissertation. The initial thought had been that I wouldn't fully appreciate the number or quality of contributions which had supported this work until it was absolutely finished. Whether inspired by procrastination or intuition, this has certainly proved to be true. With several years' worth of effort behind this final product, there are of course several sources of motivation, encouragement, and influence which have added to it. Here is an abridged review of these sources.

I am especially grateful to my advisor, John Lupton, for accepting me into his research group and acting as my mentor. Since I initially came to the University of Utah seeking a M.S. in Instrumentation Physics, and only transitioned to the Ph.D. track on a whim in order to secure a tuition waiver, it must have been quite risky for him to hire such a student. From my perspective, I have received much more than I had anticipated in this move. Where John initially encouraged my "inner engineer" for building his lab infrastructure, his excitement for scientific inquiry proved contagious and has greatly encouraged my evolution as an independent researcher. I wish to thank him for creating a research environment for his students which rewards independence and expects open exploration; weekly exposure to the scientific novelty presented at his group meetings has been terribly exciting, as well as somewhat addicting. Also, I feel that the high degree of professionalism and quality standards that is expected by him will serve me well for the remainder of my career.

Although I began my graduate work with John Lupton, I quickly became a shared student in the lab of Christoph Boehme. Becoming comfortable with the tool set of spin resonance was at first daunting, but with some effort, the power of the technique became clear and has ultimately allowed me to conduct the studies presented in this dissertation. Throughout my slow trip up this steep learning curve, Christoph has consistently encouraged my work, while simultaneously being patient with my questions and progress. From the beginning, he has given me open access to his lab space, allowed me to work on projects of my own design, and offered invaluable direction and advice. I am truly grateful to have had the opportunity to work with, and for, him.

I am also very fortunate to have intelligent, thoughtful, and quite humorous coworkers. All members of both the Lupton and Boehme group have been of the highest caliber and I consider myself lucky to have worked with them. In particular, I want to thank Nick Borys for initially disguising his intelligence, allowing me to trick myself into switching to the Ph.D. track ("because if *he* can do it then certainly *I*..."). This illusion soon unraveled to reveal my arrogance. As I began to work more closely with him, I quickly found that, in fact, he is an excellent researcher in his own right and the quality of his work is difficult to match. In much the same way, I wish to thank William Baker, who began in the Lupton–Boehme collaboration experiments with me. His internal sense of doubt has fueled many productive conversations – and I apologize for those that ended in his frustration. Additionally, the conversations, support, and camaraderie between myself and the other students and post-docs of these two groups have been highly valuable to me; I am happy to count these people among my friends.

The network of support that exists within the Physics and Astronomy Department has been of utmost help to me over the past several years. This especially includes the administrative (Jackie Hadley, Heidi Frank, Kathrine Skollingsberg, Sare Gardner), financial (Kathy Blair, Roberta McCormick, JoLene Snyder, Vicki Nielsen, Deana Young), facilities (Harold Simpson), and technical (Ed Munford, Matt DeLong, Jay Norwood) staff members. Thank you for always making my life easier and our interactions enjoyable.

I also wish to acknowledge our external collaborators, Dmitri Talapin and Jing Huang, who synthesized the high-quality nanocrystals that are the basis of my thesis work. Thank you for kindly providing me with such novel materials.

An additional special mention should be made of Nick Borys and Mark Limes, who each gave an exceptional amount of their time to proofread this manuscript. The large number of grammaktical errors they've found has been humbling.

Certainly, the seeds for pursuing a career in the sciences are planted at an early age and nurtured throughout life. In this regard, I consider myself especially fortunate. I truly thank my parents and family for sustaining an environment which placed a high value on education, experience, and fun(!). The wide ranging freedom and intrinsic trust that my parents granted me as a child have encouraged a deep sense of independence and curiosity, which I now enjoy very much. They also deserve my thanks and acknowledgment for their unfailing encouragement during my sideways trajectory through life. (*We're a ways off from pots by the road.*)

My relationship to my wife began shortly before moving to Utah to pursue this crazy impulse. In a very real way, my journey through graduate school has been her journey as well. I thank her profoundly for sharing life's adventure with me and for supporting me with the warmth of her actions, as well as the sharp honesty of her tongue.

Contents

List of Figures

Chapter 1
Introduction

1.1 Colloidal Nanocrystals

The burgeoning field of nanotechnology, in its conception, is popularly attributed to the late Dr. Richard Feynman, who first described the action of "a billion tiny factories,...drilling holes, stamping parts, and so on" in his 1959 address "There's Plenty of Room at the Bottom" to the *American Physical Society* [1]. In this spirit, he went on to propose a significant challenge to the scientific community, offering a \$1,000 reward for the construction of a 1/64th inch cube electric motor. Although met (and paid) within a year, the field of nanotechnology did not begin to develop until the early 1980s, when scanning tunneling microscopy (STM) [2] and nanofabrication [3, 4] methods began to be invented, providing tools for top-down atomic manipulation.

As the field has begun to mature, less emphasis has been placed on the simple miniaturization of traditional mechanics and more on the material aspects that dictate the quantum interactions which naturally dominate at these length scales (1–100 nm). If Feynman had anticipated that the field of nanotechnology would be so driven by materials and instead posed a challenge in these terms, he would have been tardy in payment by about 1,700 years. Incredibly, the fourth century AD Romans had developed a method of colloidally suspending gold and silver nanoparticles within glass, imparting interesting optical effects due to the surface plasmon resonance induced by light of visible wavelengths. A beautiful example of their craftsmanship survives in the Lycurgus Cup [5], the glass of which contains about 1% of roughly 70 nm diameter nanoparticles [6], giving the goblet a deep red color for transmitted light and a green color similar to copper patina for reflected light.

Knowledge-based insight into the unique light-scattering properties of metal nanoparticles only slowly emerged after an initial scientific evaluation in 1976 [7]. Colloidal glass-suspensions of *semiconducting* nanoparticles followed several years later, when in 1982, Ekimov and Onushchenko made the first size-distributed series of semiconductor quantum dots, demonstrating the quantum size effect

K. van Schooten, *Optically Active Charge Traps and Chemical Defects in Semiconducting Nanocrystals Probed by Pulsed Optically Detected Magnetic Resonance*, Springer Theses, DOI 10.1007/978-3-319-00590-4_1, © Springer International Publishing Switzerland 2013

on the electronic band gap through a correlated shift in the onset of excitonic absorption [8]. This original result of being able to continuously vary the frequency of light either absorbed or emitted from a material simply based on its size dimensions led to a great deal of early interest in nanocrystals. Since then, exerting synthetic control over the dimensional, geometric, and material aspects of quantum confinement has been realized in a wide variety of ways, ultimately revealing a broad range of novel effects and potential applications.

Somewhat traditional top-down engineering techniques are widely employed in building various quantum confined structures. High-precision electron beam lithography is used in building electrostatically defined quantum wells of interacting 2D electron gasses [9] and well-controlled reactive chemical techniques are used in the manufacture of epitaxially grown quantum dots and wires [10]. A newer generation of fabrication methods rather employs a bottom-up approach, following the Romans' earlier intuition. These range in concept from the small colloids just mentioned to the self-assembly of extended macroscale objects from nanoscale constituents [11]. The technological utility of glass-matrix colloids is ultimately limited, and so they have largely been replaced by more facile wet-chemical methods of producing solution-suspended nanoparticles [12–16] (see Sect. 2.3 for discussion). At this point, this class of nanostructure inherits the benefits of being solution-processable, making device production potentially cheap and simple through the use of conventional manufacturing processes like reel-to-reel processing [17] or inkjet printing [17, 18].

Optoelectronic device concepts such as light-emitting diodes (LEDs) [18], video displays [18, 19], photovoltaics [20–24], lasers [25], and nanomedicine [26] have all been explored using these solution-processable colloidal nanocrystals. The successful market realization of such devices ultimately depends on their efficiency, which is controlled by the stability of excitations within these active materials. Unfortunately, the ubiquitous presence of charge "trap" and crystalline defect states has helped forestall the realization of cost-competitive device manufacturing in this area. These undesirable states degrade device performance by offering competing energetic relaxation pathways [27–29] for the more desirable excitonic [24, 30] and multi-excitonic [31, 32] excited states. The effect on excitonic states is readily seen via observations of photoluminescence (PL) intermittency ("blinking") [33], power-law optical decays [30, 34], and the lack of a phonon bottleneck [35, 36].

1.1.1 Electronic Properties

Bulk semiconductor crystals (i.e. dimensions $\gg 10$ nm) are generally characterized by the electronic band gap formed from their periodic bonding structure. The equilibrium distance of the constituent atoms forces the overlap of electronic orbitals, leading to lifting of degeneracy that is driven by Pauli exclusion. Since these orbitals

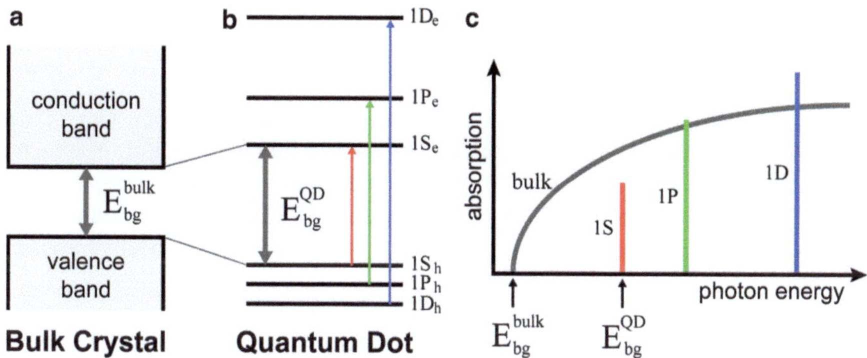

Fig. 1.1 The electronic band gap of bulk and nanoscale crystals. (**a**) Bulk semiconductor crystals are characterized by the band gap separating their continuous valence and conduction bands. (**b**) The reduced size dimensions of semiconductor quantum dots result in atomic-like energetic states. (**c**) Optical transitions between separate orbital states are discrete, reflecting the quantum dot density of states. This fact is witnessed as discernible peaks in the optical absorption spectrum, whereas the bulk material absorption is continuous in energy

delocalize across the breadth of the crystal and their number is large ($N \approx 10^{19}$), a correspondingly large number of nondegenerate orbital states forms an effectively continuous band of allowed energies. Electronic energies not supported by these interacting orbitals are disallowed, which defines the band gap of the material [37]. As the size dimensions of the bulk crystal are diminished to the nanoscale (≤ 15 nm), the number of participating valence states is also decreased. The width of each band of allowed states then begins to narrow, resulting in nearly discrete, atomic-like states for nanocrystals [38]. This is schematically depicted in Fig. 1.1.

Since the band gap of a bulk semiconductor is an intrinsic material property, much of the effort in semiconductor physics for the last 60 years has been devoted to methods of band gap engineering. Far from arbitrary, the ability to tune this intrinsic property through either doping, applying strain, or heterostructuring fundamentally enables modern electronics (e.g. diodes, transistors, LEDs, photovoltaics, etc.). This is very important to optoelectronic devices, which are based on the excitonic state, a Coulombically bound electron–hole pair that is formed at the band gap. The stability of such an excited state is due to the energy-lowering electrostatic attraction, $E_{exciton} = E_{bandgap} - E_{coulomb}$. Excitons in bulk semiconductors have an average charge carrier separation of about 10 nm (depending on the material), allowing for minimal exchange overlap of the carrier wavefunctions. This intercharge distance is termed the Bohr exciton radius since the quantized energy levels of the exciton can be treated in a Hydrogenic model [39]. Once the size dimensions of the nanocrystal are reduced to the Bohr exciton radius, strict terminal boundaries begin to be enforced for the carrier wavefunctions. This action of quantum confinement essentially creates a particle-in-a-box situation, affecting the energy-level spacing for band gap states. The band gap energy of the nanocrystal quantum dot is then [40]

$$E_{bg}^{QD} \approx E_{bg}^{bulk} + \frac{\hbar^2 \pi^2}{2\mu R^2} - \frac{1.8 e^2}{4\pi \epsilon \epsilon_o R} ,$$

where μ is the reduced electron–hole mass, ϵ is the dielectric constant of the semiconductor, and ϵ_o that of free space. R is the quantum dot radius. The factor of 1.8 in the Coulomb term arises from the Coulomb integral[1] of the $1S_e$ electron and $1S_h$ hole wavefunctions which comprise the exciton ground state. At this point, band gap engineering for the nanocrystal becomes facile and arbitrary by means of a simple choice of particle radius [41].

In combination with the continuous flexibility offered by size-tuning, the band gap can also be engineered through more traditional methods: impurity doping, strain fields, and heterostructuring. Each of these methods has been explored extensively for colloidal nanocrystal systems. For instance, doping with magnetic impurities, such as Mn^{2+} ions, has opened up possibilities for quantum dots with optically induced [42] and charge-controlled [43] magnetization. Band gap tuning through interfacial strain[44] between heterojunctions of materials with mismatched lattice constants can allow for nearly the same range of tuning available through simple particle dimensions alone. And for heterojunctions themselves, there are a wide variety of options available [45] owing to the many compatible semiconductor materials used in constructing them and the ability to synthesize a range of geometries and dimensions [45].

Heterostructure formation typically involves a layered growth procedure where one nanocrystal serves as the seed for further growth of an additional layer of another material. Since the synthetic chemist can control the material and size dimension of each growth layer, a large array of band gap configurations is possible. Both the relative band gap offset as well as the alignment of valence or conduction bands is then determined by not only the material itself, but also the size dimensions of each layer. Typically used bilayer heterojunctions are illustrated in Fig. 1.2, which shows that not only are the electron–hole energetics engineered, but wavefunction engineering is also exploited in order to tune the degree of spatial localization within the structure.

Type I band alignments generally result in both carriers being localized to the smaller band gap material, which increases wavefunction overlap and, consequently, optical recombination rates. Oftentimes, colloidal nanocrystals are terminated with a much larger band gap material, which serves as a passivation layer, protecting against the formation of dangling-bond trap states that act as localization sites and energetic decay channels. On the other hand, Type II alignment minimizes the overlap of carrier wavefunctions due to the opposing mismatch between corresponding charge carrier bands. The charge-separated states which result have a markedly decreased rate of recombination. An intermediate regime also exists, termed quasi-Type II, in which there is nearly equal alignment between the two conduction or valence bands. Such a configuration continues to enforce the localization of one

---

[1]The Coulomb integral is given by $J_{Coul.} = \int\int d\vec{r}_e d\vec{r}_h \; \Psi_e^*(\vec{r}_e) \, \Psi_h^*(\vec{r}_h) \left(\frac{1}{|\vec{r}_e - \vec{r}_h|}\right) \Psi_e(\vec{r}_e) \, \Psi_h(\vec{r}_h).$

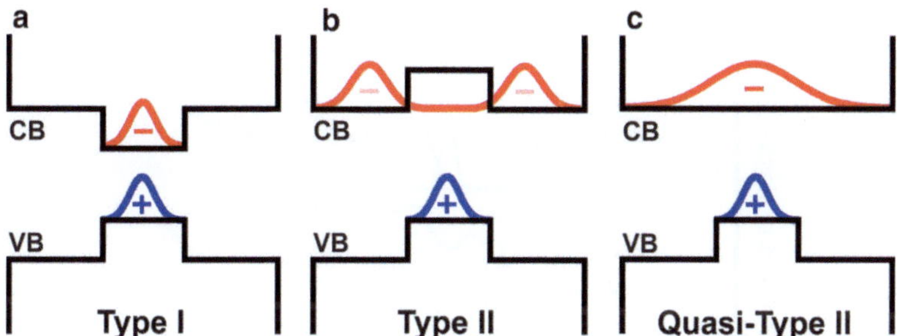

Fig. 1.2 Common types of nanocrystal heterojunctions. Three types of heterostructure are normally used for band gap tuning in semiconducting nanocrystals, which allows for charge localization via wavefunction engineering. (**a**) In Type I band structures, both electron and hole are co-localized to the same material. (**b**) Type II heterojunctions minimize wavefunction overlap by forcing charges to occupy adjacent materials. (**c**) An intermediate regime is also realizable. In these quasi-Type II heterostructures, one charge is localized while its partner is delocalized across the two materials. This effect is due to an equal alignment of either valence (VB) or conduction band (CB)

carrier, while enhancing the delocalization of the other. This ability to delocalize a single carrier has supported several studies aimed at investigating the nature of excitonic coupling in nanoscale semiconductors (i.e. via exchange [46] and electric field effects [31, 45, 47]) and the interface of their heterojunctions [48].

1.1.2   Optical Properties

As the band gap of these nanoscale materials is continuously varied, so are their optical properties. Band-edge absorption and emission frequencies directly scale with the degree of quantum confinement exerted on the exciton. With decreasing particle radii, this is observed as a corresponding optical blue-shift in both the emission and onset of absorption. A cartoon of this effect is shown in Fig. 1.3.

Although the band gap engineering described above (Sect. 1.1.1) can be useful in this regard, additional influence over the optical properties of nanocrystals is available through geometric design. By utilizing unique growth conditions and intelligently choosing particular crystalline phases and facets for heterostructure seeding, an enormous range of particle geometries can be realized. Examples include rods [15], cubes [49], pyramids [49], tetrapods [50], ribbons [51], and highly complex extended structures [52]. This geometric complexity not only furthers the ability for wavefunction engineering, but can also be used to greatly enhance the absorption cross-section. An increase in excitation rates can even be achieved while maintaining quantum confinement by using elongated rod or well structures that extend selective dimensions of the system.

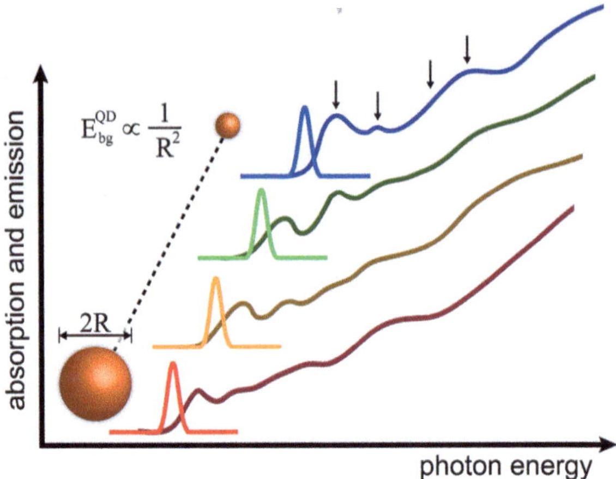

Fig. 1.3 The size-tunable optical band gap of quantum dots. Quantum confinement is enforced by material dimensions being smaller than the natural exciton length scale ($R \approx 10$ nm in bulk crystal). This results in the ability to vary the band gap as a size-tunable parameter. Therefore, smaller particle sizes display a blue-shifted optical absorption onset and emission energy. The subtle features present in the absorption spectra (*arrows*) correspond to the different orbital excitations (Data presented in this figure simulate the measured behavior)

1.1.2.1   The "Dark" Exciton

Since Chaps. 3 and 4 describe studies on spin-dependent optical states existing in CdSe and CdS nanocrystals, some attention should be paid to the angular momentum possessed by excitons in these two material systems. It should be clearly stated, though, that in these studies, the band edge exciton states are only *indirectly* addressed through the action of intermediary "trap" or "shelving" states.

As the synthesis steps for producing CdSe quantum dots of extremely high quality became developed [13], curious optical responses in these systems began to be reported in the literature. Band edge excitonic states displayed recombination lifetimes on the order of 1 µs [53, 54], whereas excitons in bulk materials radiate within roughly 1 ns [55]. An inverse relationship between nanocrystal radius and Stokes shift was commonly observed [56]. Also, a decreasing photoluminescence (PL) lifetime was demonstrated for quantum dots held in an increasing magnetic field [57]. In 1996, the matter was largely resolved in a highly revealing work by Al. L. Efros and coworkers [56], where it was shown both theoretically and experimentally that each of these separate effects could be reconciled by considering the ground state exciton of CdSe quantum dots to possess a spin-state which disallowed a radiative transition back to the ground state. Much of the subsequent work on CdSe nanocrystals has revolved around fully describing the nature of this "dark" excitonic state and how it affects the system's energetics in relation to particle size [58–60] and morphology [56, 61].

This dark excitonic state arises due to the orbital angular momentum contributions to total spin identity that are acquired by the charge carrier wavefunctions as a consequence of quantum confinement. A thorough overview of the theoretical description [56, 58, 61, 62] of these states is given by D. J. Norris in Ref. [63]. Essentially, for spherical quantum dots in the strong confinement regime (i.e. $E_{box} \gg E_{Coulomb}$ due to terms $r^{-2} \gg r^{-1}$), the electron and hole can be treated as independent particles within a spherical "box" of infinite boundaries. This allows the wavefunction for each carrier to be treated completely independent of the other. Then, each particle wavefunction can be approximated as a separable product of wavefunctions: an envelope function which satisfies the spherical particle-in-a-box potential as well as a periodic function which satisfies the crystal potential:

$$\Psi_{e,h}^{QD} \approx \Psi_{e,h}^{sphere}\, \Psi_{e,h}^{lattice}\,.$$

This separability holds as long as the nanocrystal radius is much larger that its lattice spacing, which is generally true. The total state for the electron–hole pair (ehp) is then proportional to the envelope functions of each carrier in the spherical potential, written in terms of spherical harmonics and Bessel functions,

$$\Psi_{ehp}(\vec{r}_e,\vec{r}_h) = \Psi_e(\vec{r}_e)\,\Psi_h(\vec{r}_h) \propto \left[\frac{j_{L_e}(k_{n_e,L_e}r_e)\,Y_{L_e}^{m_e}}{r_e}\right]\left[\frac{j_{L_h}(k_{n_h,L_h}r_h)\,Y_{L_h}^{m_h}}{r_h}\right].$$

Here, $j_L(k_{n,L}r)$ is an Lth order spherical Bessel function, where $k_{n,L}$ is the nth zero of j_L, and Y_L^m is a spherical harmonic. Electronic states are then labeled by the quantum numbers n_e, L_e, n_h, and L_h, reminiscent of atomic-like orbitals with $n = (1,2,3\ldots)$ and $L = (S,P,D\ldots)$. It then becomes clear why quantum dots are commonly referred to as "artificial atoms."

The full set of quantum numbers describing the total angular momentum for an exciton in a quantum dot is then six-fold. This is because, in addition to the atomic-like states just shown, there is still the orbital momentum within the atomic basis ($\ell_{e,h}$) as well as the intrinsic spin ($S_{e,h}$) that the charges possess. Each charge carrier then has a total angular momentum of $F_{e,h} = L_{e,h} + J_{e,h}$, where $J_{e,h} = \ell_{e,h} + S_{e,h}$. The exciton's good quantum number is then $N = F_e + F_h$. This composition is schematically depicted in Fig. 1.4a.

In truth, the accurate computation of carrier states for actual particles is a highly nontrivial process. Simply changing the nanoparticle radius can radically alter the relative spacing of energetic sub-bands [58, 61, 64]. This effect is due to the perturbation strength of each level upon its neighbor-levels and is more of an issue with hole states than those of the electrons since valence sub-bands (light hole, heavy hole, and spin-orbit split-off) tend to lie so close to one another. The small amount of energy separation for these states leads to band mixing, reducing the purity of each state's wavefunction. Computing the wavefunction in materials like CdSe offers some relief since the crystal-field splitting energy is so much weaker than its intrinsic spin-orbit coupling ($\Delta_{CF} = 25\,\mathrm{meV}$ and $\Delta_{SO} = 420\,\mathrm{meV}$) [64]. Mixing from the spin-orbit band then becomes negligible, allowing for a more

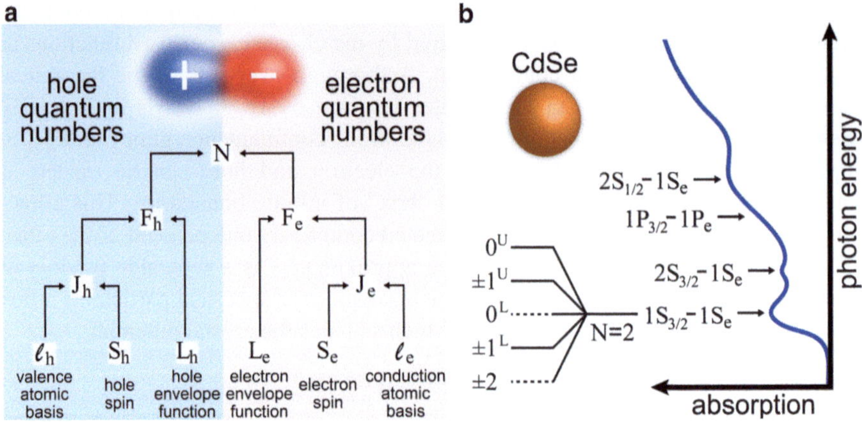

Fig. 1.4 The total angular momentum of quantum dot excitons. (**a**) Angular momentum for a quantum dot exciton is primarily composed of three quantum numbers per charge. Quantum confinement of both electron and hole results in an envelope function component to its wavefunction, gaining orbital momentum, $L_{e,h}$, for the carrier. Traditional quantum numbers, $\ell_{e,h}$ and $S_{e,h}$, continue to characterize the charges in their atomic basis. (**b**) An optical absorption spectrum for CdSe nanocrystals is shown. Individual absorption peaks are labeled according to their orbital transitions, nL_F, for each electron and hole state. The nondegenerate fine structure for the ground state exciton is displayed. *Solid lines* indicate allowed ("bright") optical transitions, while *dashed lines* are optically forbidden ("dark") excitonic transitions (Data presented in this figure simulate the measured behavior)

approachable theoretical treatment [56, 62]. Realistically computing correct holeband energy levels in materials with nearly equal spin-orbit and crystal field energies, $\Delta_{SO} \approx \Delta_{CF}$, on the other hand, has proved challenging for theorists, with an advanced treatment for CdS only being developed in 2010 [62].

The ordering of hole levels in nanocrystals is highly important since it determines the final spin-identity of the excitonic state. For both CdSe and CdS [56, 62], the lowest energy electron and hole states are generally found to be $\left[1S^e, 1S^h_{3/2}\right]$ (where $F_h = 3/2$). Since $F_e = 1/2$, the total exciton spin identity has eight states, three of which are degenerate in absence of Zeeman splitting. Ordered by decreasing energy, these are 0^U, $\pm1^U$, 0^L, $\pm1^L$, ±2, where the superscripts "U" and "L" stand for "upper" and "lower," respectively (see Fig. 1.4b). Here it is seen that the ground state exciton is a spin-2 state. Spin relaxation between $N = \pm1$ (optically "allowed") and $N = \pm2$ (optically "forbidden") levels has been reported to be very efficient ($T_1 \approx 1\,\text{ps}$) for CdSe excitonic states [65]. Since there is no electric-dipole-allowed transition from $N = \pm2$ to the ground state, energy relaxation occurs over very long time scales, even at low temperatures. This explains the long $\tau \approx 1\,\mu\text{s}$ fluorescence lifetimes observed for CdSe nanocrystals at $10\,\text{K}$ [53, 54], which were mentioned above.

As one would expect from the above discussion of hole-band ordering as a function of particle radius, the relative energetic spacing of the excitonic fine

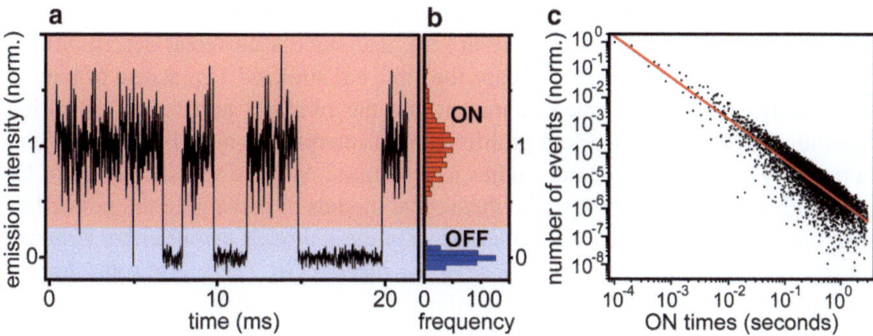

Fig. 1.5 Characteristics of single particle fluorescence intermittency. (**a**) Single particle blinking is characterized by stochastic switching between optically bright and dark states. (**b**) A bimodal distribution of emission intensity counts shows a clear delineation between these bright (ON) and dark (OFF) emission intensity levels. (**c**) Analysis of long-term blinking traces reveals a power-law distribution in the amount of time spent in an ON or OFF state. This behavior confirms the complex dynamics underpinning these random switching events (Data presented in this figure simulate the measured behavior)

structure, and even the ordering itself, is a strong function of nanocrystal size and geometry [58, 61, 64]. Recently, wavefunction engineering in quasi-Type II CdSe/CdS core-shell nanocrystals has resulted in a dramatically reduced exchange interaction, decreasing the roughly 15 meV gap between ± 2 and $\pm 1^{L}$ levels to less than 250 µeV [46]. Thermal fluctuations were then sufficient to effectively depopulate the "dark" exciton state, resulting in enhanced emission stability.

1.1.2.2 Fluorescence Intermittency ("Blinking")

Soon after methods for synthesizing colloidal nanocrystals of high quality became known, a surprising characteristic of the emission process was discovered. It was found that, under constant excitation conditions, single nanocrystals display a rather binary emission intensity as a function of time, where they seem to be either emissive or not [66]. This fluorescence intermittency of the particle (a.k.a. "blinking") shows stochastic switching behavior between the two primary emission states, being in either an "ON" (bright) or an "OFF" (dark) state; the process highly resembles the form of random telegraph noise [33] (see Fig. 1.5a, b). Like the action of random telegraph noise, the distribution of dwell times for the emitter while in either the ON or OFF state has a power-law distribution [67] (Fig. 1.5c), meaning that there is no average ON time for the emitter.

Random blinking behavior is not uncommon amongst quantum emitters. In fact, it is a well-known and well-described process in ionic [68] and molecular [69] systems where the observed dynamics can be attributed to competing energy decay mechanisms. This usually takes the form of a short-lived optically allowed transition competing with a lower-lying optically forbidden state with a long lifetime.

The mechanism underlying the blinking process in colloidal nanocrystals has proven to be a much more complicated affair [70], and since a universal description has remained elusive for the past 16 years, the topic has attracted a great deal of interest. The many studies performed characterizing the blinking process under various conditions [71–74] has led to a proliferation of interpretive models [33, 75]. Since a majority of the experimental studies have focused on the stochastic nature of ON and OFF dwell times, so have the theoretical models in explaining the power law-like distributions of these dwell times. Most of these models invoke either a random charging event leading to a persistent Auger process [76, 77] or else some form of charge localization [78, 79], whether to traps, defects, or even the surrounding matrix material. Discrimination between these models has been notoriously difficult for this material system, though some fraction of resolution has recently been offered with the insight that *both* Auger and trapping processes can induce a blinking event [27–29]. What remains completely ambiguous at this point, though, is what the causative factors are for these processes; how are single charges introduced into the nanocrystal and what particular chemical defect and charge trapping sites are responsible?

Understandably, a great deal of effort has been directed towards eliminating this intermittent behavior from this class of materials. One very effective approach is to engineer the wavefunction of the system in order to minimize its overlap with any defect or trapping states (presumed to consist of dangling bonds at the terminal surface) [80]. This is accomplished by exchanging an abrupt core-shell hetero-junction for a continuous band transition via radial-alloying. Surprisingly, blinking is completely suppressed in this case, despite the particle undergoing the same charging events normally associated with Auger-induced OFF states. Since this technique only applies to core-shell heterojunctions of specific materials, a second major approach to enhanced optical performance has been widely explored. This is through the engineering of more effective surface passivation techniques [22, 23]. Broadly employed, organic ligands are used to satisfy the terminal bonds of these nanocrystals, and therefore, the particle surface is densely populated with a layer of these materials. Blinking dynamics have also been explored in terms of passivation type [73, 81] and matrix environment [82], where it has become obvious that the particle's surface states also play a role, again, as a source of trap states. Results from these studies highlight the need for a chemically complete understanding of the physical mechanisms involved, both in driving the blinking process and in quenching it. Such information could then be used to eliminate any chemical or energetic support for the production of OFF states.

1.1.3 Trap and Defect Properties

Common to the study of crystalline materials is the subject of charged "defect" states. In bulk materials, these are generally formed from some sort of interruption to the lattice periodicity in either one (point defects), two (line defects), or three dimensions (bulk defects). For nanocrystals, the sheer lack of particle volume restricts the type to point defect centers (i.e. vacancies, interstitials, clusters, etc.). Furthermore, since these systems are characterized by their very large surface-to-volume ratios, unsatisfied and disordered bonding at the surface can potentially dominate the energetics of the entire system [30]. This poses a problem of utmost importance for optoelectronic device applications since these states provide strong electronic decay pathways [83], decreasing the efficiencies of LEDs [84], solar cells [24], and laser gain media [85].

The inevitable existence of dangling bonds at the terminal surface can be mediated, though, by somehow satisfying these unpaired orbitals. Two common methods of accomplishing this are by covalently bonding either organic ligands or a capping shell of wide band gap material to the surface. Unfortunately, neither of these methods has been found to function completely against surface defect states. In fact, it has been shown that passivating ligands themselves can introduce energetic trap states by being incorrectly bonded to the surface or by having a suboptimal packing ratio [86, 87].

In nanocrystals, the broad classifying terms of "trap" and "defect" somewhat overlap, and are both ill-defined, as there is very little chemical knowledge available for these states. In general, though, traps are nearly always associated with shallow or metastable states, possibly induced by environmental conditions (e.g. ligands or the host matrix), while defects refer to classical crystallographic discontinuities. The overlap in terminology comes from the wide variety of deformations which occur at crystalline surfaces reconstructed by competing ligand orbitals [86]; even for bulk semiconductors, descriptions for many surface states remain undefined or imprecise due to the complex atomic reorganization which normally takes place at surfaces [88]. In any case, some work has been applied to characterizing these states in order to quantify their detriment to device energetics [24, 30] and in hopes of gaining the chemically relevant information [89, 90] needed in order to minimize these states during the synthesis process.

Since spectroscopy of emission dynamics serves as a direct window on excitonic states and their perturbations, photoluminescence studies have been fairly powerful tools in trap state investigations. Spectral time dynamics resolved on the 100 fs time scale indicate that carrier trapping in CdS nanocrystals can be as fast as a few picoseconds [91]. Indeed, a recent theoretical framework has been developed to study trapping kinetics as a function of band gap engineering and trap depth, affirming the femto − to nanoscale trapping times [79]. On the other hand, the *trapping lifetime* can be quite long, ranging from nanoseconds to several microseconds [71, 92], as demonstrated, again, in both fluorescence decay measurements and model calculations [30]. Section 3.6.2 of this work has some further discussion on deciphering the presence of trap states from the form of emission decay dynamics.

Generally, trap and defect sites are presumed to represent nonradiative decay pathways for band edge excitonic states. A particular species of deep-energy chemical defect, though, is actually emissive and so represents an additional radiative decay channel for excitonic states [90]. Common to both CdS and CdSe materials, and largely the subject of Chap. 4, this emissive site has been studied for more than 60 years [93]. Despite the history of work, little is known about the exact chemical or structural nature of this site, other than that it is probably a vacancy defect cluster [89, 94]. In CdSe and CdS, this would effectively be a donor-acceptor pair formed from a Cd vacancy (hole trap) located nearby either a Se or S vacancy (electron trap) [95]. What *is* known is that its emission is a multiphonon-driven process, as is evident from its wide emission band [96], lack of a visible 0–0 transition, and therefore large Huang-Rhys factor ($S = 18$) [95]. The emission spectrum for this emissive defect in CdS nanorods is shown in Chaps. 3 and 4, where the spin-dependencies of this recombination channel are discussed.

Aside from optical forms of defect spectroscopy, several other techniques have been applied to the characterization of these nebulous trap and defect states [97]. Several studies relying on cyclic voltammetry have been made [98]. This technique holds the material of interest in an electrolytic solution while the electrochemical potential is cycled. Current is monitored and correlated with potential in order to observe detrapping events, which correspond to spiking in current. A newly developed variation of this technique is to instead monitor the emission spectrum of the material of interest while sweeping the electrochemical potential of this cell [29, 99]. Since doing so shifts the Fermi level of the material, control over the trap state occupancy can be made, revealing the effect these states have on emissive band edge and defect sites. Traditional electronic probes can also be revealing: conducting atomic-force microscopy (C-AFM) addresses these sites on the atomic scale [100]; deep level transient spectroscopy (DLTS) helps determine trap concentrations [101]; and photoconductivity measurements probe trap types through below-gap [102] and field-dependent effects [103].

Finally, the technique of continuous wave, optically detected magnetic resonance (cwODMR) has proved to be fairly powerful in its ability to directly address long-lived trap charges [94, 104]. Virtually all work involving this technique as applied to colloidal nanocrystals has been performed by, or at least involved, Dr. Efrat Lifshitz of the Israel Institute of Technology – Technion [105–114]. The power of using a pulsed form of this method (pODMR) in characterizing a site's electronic and chemical environment is the topic of Sect. 1.2, as well as of the studies presented in Chaps. 3 and 4.

1.2   Pulsed Optically Detected Magnetic Resonance (pODMR)

1.2.1   Electron Spin Resonance (ESR) of Optically Active Carriers

Electron spin resonance is inherently based on the interactions of an electronic charge's intrinsic magnetic moment with an external magnetic field, where the moment either aligns parallel or antiparallel to the field. The relative difference in energy between these two alignment configurations is called Zeeman splitting, after Pieter Zeeman who in 1896 first observed this behavior in atomic emission lines[115], well before the quantum nature of the effect could be appreciated. Since this interaction scales linearly with magnetic field,[2]

$$\mathcal{H}_Z = -\boldsymbol{\mu} \cdot \mathbf{B}_0 = g\beta_e \mathbf{S} \cdot \mathbf{B}_0 ,$$

the magnitude of emission line splitting was fundamentally instrumental to the remote measure of sun-spot magnetic field strengths [116] and polarizations [117]. It was in 1944, though, that the Russian physicist E. K. Zavoisky used the Zeeman splitting of electronic spin states to invent ESR as a chemically specific spectroscopic technique[118]. The technique quickly found utility with chemists in the study of free radical generation in salts [119, 120] and organic crystals [121], as wells as point defect structure in inorganic crystals [122–124]. More recently, the framework of ESR has been of paramount utility to physicists working on the purposeful manipulation of quantum states [125–127] and moving the field towards the realization of "qubit"-based quantum computers [128–130].

The polarization of large ensembles ($N_{spins} \geq 10^{10}$) is what traditionally enables the observation of ESR signals through radio and microwave frequency absorption. Several variations of spin readout method have been devised since the inception of ESR, increasing the sensitivity and specificity of the technique. Readout methods based on relative spin permutation symmetry, rather than simply spin polarization, allow for resonance detection even at the level of single spins [127, 131]. Optically detected magnetic resonance (ODMR) is such a method, where ESR of optically active carriers is observed through either the fluorescence [132] or phosphorescence [133] emission channel.

Since optically active carriers are intrinsically paired (i.e. excitons, polaron-pairs, etc.), the mutual spin configuration between electron and hole is what defines the orthogonal states of the two-spin system. In the absence of additional angular momentum, this relative spin orientation is defined to be either a singlet ($\mathbf{S} = 0$) or triplet ($\mathbf{S} = 1$) state. Upon gaining additional angular momentum, either through atomic orbital motion or quantum confinement effects [57], the spin multiplicity of the excitation can be increased (see Sect. 1.1.2.1 for additional discussion). The oscillator strength for an optical transition out of any one of the excitation's spin states is then governed by spin selection rules. For example, in most organic semiconductors, the excited singlet state has a high oscillator strength and so optical transitions (fluorescence) are very efficient for this configuration. Alternatively, the direct transition of an excited triplet state to the singlet ground state in organics is dipole–forbidden, producing only very weak optical emission (phosphorescence) and usually at lower energies with long decay lifetimes [134, 135]. Thus, there

---

[2]Here, $\boldsymbol{\mu}$ is the magnetic moment of the carrier, $\mathbf{B}_0$ is the external magnetic field, g is the Landé g-factor, $\mathbf{S}$ is the spin angular momentum of the charge, and $\hbar$ is Planck's constant. The Bohr magneton, $\beta_e = \frac{e\hbar}{2m_e}$, is also used, where e and m_e represent the electron charge and mass, respectively.

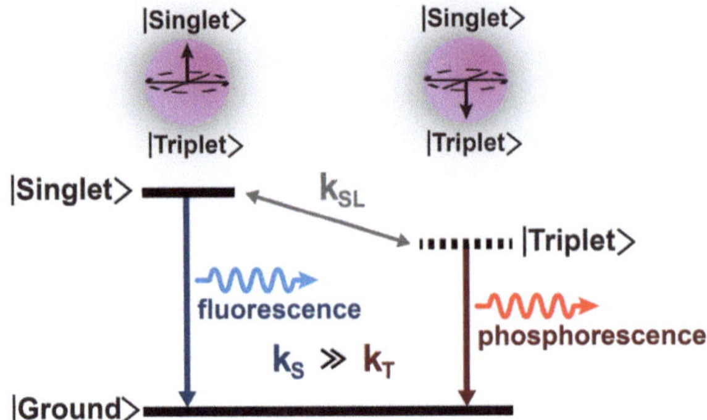

Fig. 1.6 Spin selection rules govern optical emission. The probability of an optical transition between ground and excited states depends on whether angular momentum is conserved in the transition. For a pair of spin-$\frac{1}{2}$ carriers in a singlet ground state, dipole–allowed optical transitions to (absorption) and from (fluorescence) an excited state are efficient. Some finite rate of intersystem crossing due to spin-lattice relaxation (k_{SL}) results in a triplet state population. Since transitions to the ground state are dipole–forbidden for the triplet manifold, the emission lifetime of phosphorescence is much longer than for fluorescence (i.e. $k_S \gg k_T$). Thus, for microwave-induced spin mixing that is fast compared to k_{SL}, the intensity of each emission band can be used as the observable in a spin resonance experiment involving optically active carriers

is a direct correlation between mutual spin identity and the observable of either fluorescence or phosphorescence intensity (see Fig. 1.6). For weakly exchange coupled states, the wavefunction, Ψ, always carries some superposition of singlet and triplet states. In this case, the observable is then proportional to the *amount* of singlet content shared between carriers [136], $PL \propto \left| \langle Singlet | \Psi \rangle \right|^2$.

In general, there are two methods of applying radio-frequency (i.e. the $\mathbf{B}_1$ field) in magnetic resonance; either spin excitation is continuously driven (cw), or it is driven on a time scale which is short compared to the dephasing time, T_2, of the spin (pulsed). Investigations using cwODMR produce a spin resonance spectrum, which, in principle, reflects the environment and interactions that the spin system experiences (see Sect. 1.2.2.2). In practice, though, the effects of separate interactions become convoluted in the cw spectrum, making some parameters impossible to extract. Such dependencies include, but are not limited to, the density of spins in the sample, the number of resonant species, the resonator quality factor, lock-in detection settings, g-factor or hyperfine anisotropies, dipolar interactions, and field inhomogeneities. Most of these issues can be resolved with information gained from the dynamics of the spin system. In pODMR, several pulse sequences [133] have been designed in order to probe the various dynamics and interactions which constitute the environment of the isolated spin. In this way, a different pulse sequence may be engineered in order to access one set of interactions while minimizing any convolution with other, unconnected interactions.

The remaining sections of this chapter are devoted to first describing some of the interactions which form the cwODMR spectrum and then the coherence effects which can be used to access particular aspects of the paramagnetic center and its environment.

1.2.2 Resonance Structure

1.2.2.1 Spin-Orbit Coupling and the Landé g-Factor

The fingerprint of a particular resonance spectrum is contained within the conditions required to achieve that resonance. For ESR, the resonance condition is met when an oscillating $\mathbf{B}_1$ field matches the Zeeman splitting in energy for an electronic state,

$$h\nu_{RF} = 2g\beta_e\, \mathbf{S}\cdot\mathbf{B}_0 \,,$$

where ν_{RF} is some resonant radio frequency providing $\mathbf{B}_1$ perpendicular to $\mathbf{B}_0$. The Landé g-factor here serves as a correction factor for the charge's magnetic moment. Spin contributions in the Dirac equation place the free-electron g-factor at exactly $g_e = 2$, although vacuum fluctuations in quantum electrodynamics (QED) predict a slightly larger value, $g_e = 2.0023\ldots$ [137]. Since the QED treatment for g-factor is directly related to the fine-structure constant, α, highly precise measures of g_e [138] have enabled correspondingly high precision in α [139]. To date, g_e is one of the most precisely measured quantities of science (0.76 part per trillion uncertainty [138]).

For electrons which are *not* free (i.e. atomically bound charges), the g-factor can take on a quite different value due to its orbital motion, which can serve as useful information about the intrinsic nature of the paramagnetic site. The effect fundamentally arises from the coupling of orbital ($\boldsymbol{\ell}$) and spin ($\mathbf{S}$) angular momentum. Since this is a magnetic interaction of the field generated by the orbital motion of the charge and the spin of that charge, this ultimately affects the magnetic moment of the carrier, shifting the correction factor away from g_e. This behavior represents a perturbation to the spin system, $\mathcal{H}_{SO} = \lambda\boldsymbol{\ell}\cdot\mathbf{S}$, which effectively admixes excited and ground state wavefunctions, as according to perturbation theory [140],

$$|0\rangle \longrightarrow |0\rangle + \sum_{n\neq 0} \frac{\langle n|\,\lambda\boldsymbol{\ell}\cdot\mathbf{S}\,|0\rangle\,|n\rangle}{E_n - E_0}\,.$$

Here $|0\rangle$ and $|n\rangle$ are the ground and excited state wavefunctions, respectively, with E_0 and E_n being their corresponding energies. The factor λ in the spin-orbit Hamiltonian determines the strength of the interaction, as well as the directional shift away from g_e; this goes as [141, 142]

$$g_{ij} = g_e + 2\lambda \sum_{n\neq 0} \frac{\langle 0|\,\ell_i\,|n\rangle\,\langle n|\,\ell_j\,|0\rangle}{E_n - E_0}\,, \quad i,j = x,y,z\,.$$

The magnitude of λ depends on the orbital being occupied and is proportional to Z^4, Z being the atomic number. Obviously, the magnitude of mixing also depends on energy level separation – a point which is very relevant to the mixing of hole band states of quantum dots, as mentioned in Sect. 1.1.2.1. The sign of λ depends on the cumulative spin-orbit contributions from empty (negative sign) and full (positive sign) molecular orbitals. In this way, it is possible to have large deviations from g_e [143] of both positive and negative values [144].

The g-factor can then be used as a very effective spectroscopic marker for identifying specific paramagnetic centers since it is defined by a unique combination of atomic number, occupied orbital, and electronic band structure. The precise g-values of many radicals [145, 146], defects [147], and dopants [148, 149] have been well characterized, allowing for their presence and density within a material to be ascertained. By taking advantage of the inherent stability of most paramagnetic systems, calibration-free magnetic field sensors can be produced [150, 151]. Unambiguous access to g-factor can sometimes be complicated, though. In general, the g-factor is in fact a $\mathbf{g}$-tensor, due to the anisotropic distribution of spin-orbit interactions within the local environment of a spin center. This results in multiple peaks being present in the resonance spectrum of randomly oriented ensembles, with each peak corresponding to the principal values of the $\mathbf{g}$-tensor (see further discussion in Sect. 3.6.7). Aside from this convolution effect, there also exist several mechanisms that alter or distort the resonance lineshape, which is the topic of the next section.

1.2.2.2 Broadening Mechanisms

The intrinsic lineshape of fully isolated and isotropic spin centers is a single peaked Lorentzian profile, as shown in Fig. 1.7a. The width of this line is determined by the spin dephasing time and the area is proportional to the paramagnetic number density. These parameters, along with g-factor, are sufficient information for a great deal of studies. But since ESR samples are normally measured in bulk, or in very high number, any local variations in the radical environment throughout the material ensemble will result in a Gaussian distribution of single resonance lines (Fig. 1.7b). This distribution of local environments can have several intrinsic sources (e.g. hyperfine, dipolar, etc.), but each type of interaction essentially has the same effect on the immediate environment of the resonant spin: it generates a local magnetic field, $\mathbf{B}_{local}$, which perturbs the externally applied field, $\mathbf{B}_{total} = \mathbf{B}_0 + \mathbf{B}_{local}$.

A complete description of each interaction will not be given here since a full treatment of each of these broadening mechanisms is usually found in any spin resonance text [152]. Essentially, though, these interactions fall under two general classes: (1) those that directly generate local fields; and (2) those that modify local fields.

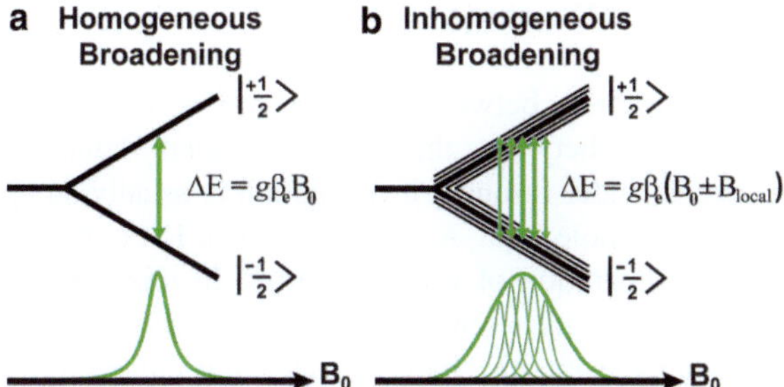

Fig. 1.7 Spin resonance lineshape broadening. Monitoring transitions between Zeeman split energy levels of an ensemble spin-$\frac{1}{2}$ system results in a characteristic resonance lineshape. (**a**) When the local environment is identical for each member of the ensemble, the lineshape is a single Lorentzian and is homogeneously broadened only by the intrinsic linewidth of the transition. (**b**) If, however, the spin ensemble experiences some normal distribution of local field perturbations, then the resonance condition for each member will also follow the normal distribution. The envelope of individual Lorentzian lines takes on a Gaussian profile with an inhomogeneously broadened linewidth

Many nuclear isotopes carry their own intrinsic magnetic moment, **I**, which can directly introduce a magnetic field offset for the electron (i.e. hyperfine interaction). Performing isotope exchange or isotopic purification, when practical, can then be a powerful method of probing local nuclear interactions [153–155]. Similarly, nearby electronic spins can couple to a paramagnetic center through a magnetic dipole-dipole interaction. This allows for intercharge distance measurements of up to 8 nm [156] to be made – a powerful technique with which to probe protein dynamics [157], for example.

Effects which simply modify the local field of a paramagnetic center normally do so by affecting the spin-orbit coupling of that center. As mentioned in the previous section, *g*-factor anisotropy arises from anisotropic spin-orbit coupling. It is reasonable, then, that any interaction which modifies spin-orbit coupling will in turn act as a local magnetic perturbation. So, since spin-orbit sensitively depends on details of the atomic orbital, any adjustment to wavefunction distributions will result in a shift of spin-orbit coupling strength. This permits electric field effects, like the internal crystal field [124] or even an external electric field [158], to lift the orbital degeneracy of a state. Additionally, a nuclear moment $\mathbf{I} \geq 1$ indicates a nonspherical charge distribution among the constituent nucleons. This creates an electric nuclear quadrupole moment which then interacts with the surrounding electronic and ionic charge distribution [159], effectively acting as a strain field [160]. In fact, when strain is present across the crystalline or molecular system, it usually results in significant broadening. Since strain modifies bonding lengths and angles, there can be a large distribution of strengths represented for hyperfine, dipole-dipole, nuclear quadrupole, crystal field, and spin-orbit interactions [161].

1.2.2.3 The Half-Field Resonance

When the physical distance between two carriers is small ($\leq 8\,$nm), magnetic dipole-dipole alignment between the two constituent spins can take place. In ODMR, the species under resonant investigation is usually an optically excited, closely bound, electron–hole pair. As stated in Sect. 1.1.1, bulk crystal excitons have a mean charge separation of roughly $10\,$nm. In this case, dipolar coupling is considered weak since its strength goes as r^{-3}. For the case of quantum confinement, though, the excitonic charge separation is enforced by the nanocrystal boundaries, allowing for a mean distance of even a few nanometers. In this range, dipolar coupling is greatly enhanced and exchange coupling becomes pronounced due to the increased wavefunction overlap. For a pair of spin-$\frac{1}{2}$ carriers, the total spin angular momentum of the system then becomes either singlet ($\mathbf{S} = 0$) or triplet ($\mathbf{S} = 1$), where degeneracy in the triplet states is lifted by the dipolar interaction. In general, ESR of spin-1 states presents unique features to the resonance spectrum, which are considered below.

The Hamiltonian for the dipole-dipole interaction between constituent spins $\mathbf{S}_1$ and $\mathbf{S}_2$ is

$$\mathcal{H}_{dip} = \frac{\mu_0}{4\pi} g_1 g_2 \beta_e^2 \left[\frac{\mathbf{S}_1 \cdot \mathbf{S}_2}{r^3} - \frac{3\,(\mathbf{S}_1 \cdot \mathbf{r})\,(\mathbf{S}_2 \cdot \mathbf{r})}{r^5} \right] = \mathbf{S}_1 \cdot \mathbf{D} \cdot \mathbf{S}_2 \, ,$$

where $\mathbf{D}$ is the dipolar tensor, sometimes called the zero-field splitting (ZFS) tensor, with elements

$$D_{ij} = \frac{\mu_0}{8\pi} g_1 g_2 \beta_e^2 \left\langle \frac{r^2 \delta_{ij} - 3ij}{r^5} \right\rangle \, , \quad i, j = x, y, z.$$

In the principal-axis system of $\mathbf{D}$, the Hamiltonian becomes

$$\mathcal{H}_{dip} = D_{xx} \hat{\mathbf{S}}_x^2 + D_{yy} \hat{\mathbf{S}}_y^2 + D_{zz} \hat{\mathbf{S}}_z^2,$$

which can be parametrized in terms of two factors,

$$D = \frac{3}{2} D_{zz} \, , \quad E = \frac{1}{2} \left(D_{xx} - D_{yy} \right),$$

allowing the Hamiltonian to be rewritten as [162, 163]

$$\mathcal{H}_{dip} = D \left[\hat{\mathbf{S}}_z^2 - \frac{1}{3} S(S+1) \right] + E \left(\hat{\mathbf{S}}_x^2 - \hat{\mathbf{S}}_y^2 \right).$$

The geometric symmetry of the paramagnetic site plays a large role in determining the $\mathbf{D}$-matrix elements. For sites with axial symmetry, $D_{xx} = D_{yy}$, resulting in $E = 0$. Lower symmetry sites (i.e. rhombic) in general have $D_{xx} \neq D_{yy}$ and so $E \neq 0$. Dipolar interactions lead to resonance lineshapes with unique profiles, determined

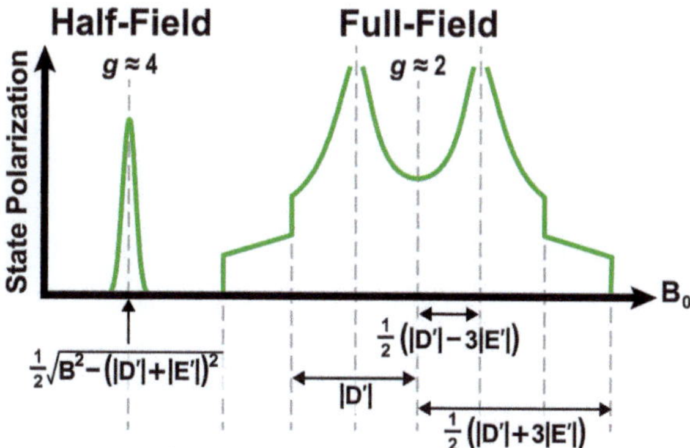

Fig. 1.8 Extracting dipolar coupling parameters from the resonance lineshape. In the presence of electron–electron dipolar coupling, the full-field resonance has a characteristic lineshape determined by the D and E parameters of the dipolar Hamiltonian. Here, D and E are in units of magnetic field ($D' = D/g\beta_e$ and $E' = E/g\beta_e$). An additional resonance is also observed at $g \approx 4$, or at about half the magnetic field of the $g \approx 2$ resonance

by the distribution of spin–spin orientations within a sample. For a system of randomly oriented triplet centers (with respect to the external field), this results in a Pake-doublet lineshape distribution [164]. Schematically shown in Fig. 1.8, this structure allows the characteristic D and E terms to be extracted [165] from which the pair separation can be calculated [166].

In cases where the dipolar lineshape features of the full-field ($g \approx 2$) resonance are absent due to poor resolution, additional restrictions can be made on the D and E parameters by relying on the observation of the "half-field" resonance. This ESR transition occurs for spin pairs which have a total spin angular momentum of $\mathbf{S} \geq 1$, and resides at about half the field strength of the full-field resonance; so at $g \approx 4$ if the full-field is observed at $g \approx 2$. The $g \approx 4$ feature is sometimes referred to as a "double quantum" transition since it appears to involve a double change of angular momentum, $|\Delta m_s| = 2$ (see Figs. 1.8 and 1.9). This potential source of confusion can be explained by the nature of the triplet wavefunction in the high and low field limits.

X-band ESR probes the $g \approx 2$ state at larger magnetic fields where the Zeeman split triplet sublevels are well defined as $|m_s\rangle = |+1\rangle$, $|0\rangle$, and $|-1\rangle$. At the much lower fields where $g \approx 4$ is observed, the purity of the triplet sublevels is diminished due to a lack of an external axis of quantization (i.e. B_z). Instead, quantization is along the internal molecular frame, resulting in a superposition of full-field triplet states:

$$|T_x\rangle = \frac{1}{\sqrt{2}}(|-1\rangle - |+1\rangle), \ \left|T_y\right\rangle = \frac{i}{\sqrt{2}}(|-1\rangle + |+1\rangle), \ \text{and} \ |T_z\rangle = |0\rangle.$$

So, in fact, transitions between the low-field $|T_x\rangle$ and $\left|T_y\right\rangle$ triplet lines are not 2-photon transitions at all, but are actually energy matching 1-photon transitions

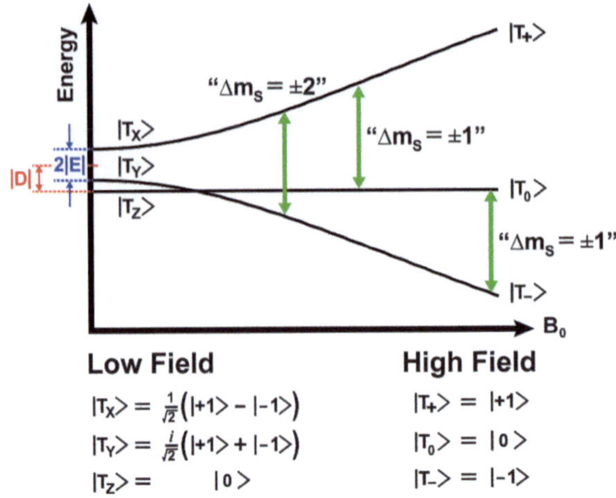

Fig. 1.9 The zero field splitting of triplet states. For a spin-1 system, there are three Zeeman split sublevels. At high magnetic fields, the system has a well-defined quantization axis (e.g. B_z). However, at low magnetic fields, the external field and internal dipolar fields mix, causing m_s to cease to be a good quantum number. With zero external field applied, triplet degeneracy is partially lifted by the internal dipolar field, resulting in a level splitting which is determined by the D and E parameters of the dipolar Hamiltonian. Wavefunctions at low external field are a superposition of full-field states, allowing seemingly spin–forbidden transitions to occur

which effectively induce a momentum change of $|\Delta m_s| = 0$ [167]. Double quantum transitions can be induced, in principal, though this is a second-order effect, and so high-intensity resonant fields are required [168].

Although observing the half-field resonance generally allows for uncertainty minimization in the measured values of D and E parameters, there are a few instances of convolution which can obscure such information. The first arises from the existence of multiple full-field resonances with significantly different g-factors. Since the resonance $\mathbf{B}_0$-field is used to determine D and E [165], it becomes impossible to choose the correct resonance for doing so, especially if each center experiences some amount of dipolar coupling. An example of such a situation occurring is given in Sect. 4.6.

The second situation which can obscure dipolar information involves the existence of significant spin-orbit interaction. Somewhat unfortunately, when spin-orbit is combined with the angular momentum Hamiltonian, $\mathscr{H} = \beta_B \mathbf{B}_0 \cdot \left(\boldsymbol{\ell} + g\mathbf{S}\right) + \lambda\boldsymbol{\ell} \cdot \mathbf{S}$, a term that is identical in form to the dipolar interaction can be produced, $\mathbf{S}_1 \cdot \mathbf{D}^{SO} \cdot \mathbf{S}_2$. This new dipolar contribution can be parametrized in terms of its own set of D^{SO} and E^{SO} parameters, which convolve with the D and E parameters from the magnetic dipolar interaction. Fortunately, since spin-orbit coupling determines the elements of the g-tensor (as seen in Sect. 1.2.2.1), its contribution to the $\mathbf{D}$-tensor takes on a similar form:

$$D_{ij}^{SO} = \lambda^2 \sum_{n \neq 0} \frac{\langle 0 | \ell_i | n \rangle \langle n | \ell_j | 0 \rangle}{E_n - E_0} , \quad i, j = x, y, z.$$

So, by having complete experimental information on the principal values of the g-tensor, the D^{SO} and E^{SO} parameters can be computed by

$$D^{SO} = \frac{\lambda}{2} \left[g_{zz} - \frac{1}{2} \left(g_{xx} + g_{yy} \right) \right] , \quad E^{SO} = \frac{\lambda}{2} \left(g_{xx} - g_{yy} \right).$$

At this point, the difference in experimentally obtained D and E parameters from those calculated through the g-tensor will reflect the true magnetic dipolar contribution [162].

1.2.3 Coherence Effects

1.2.3.1 Spin–Rabi Oscillations

One of the hallmarks of coherence in a two-level system is the observation of a Rabi oscillation. First described by I. I. Rabi in 1937 for nuclear magnetic resonance (NMR)[169], this is a coherent cycling of a system between two of its nondegenerate eigenstates, φ_1 and φ_2. In principle, any two-level system can support such resonant driving, and so this action has been confirmed for many physical systems (e.g. NMR [170], ESR [133], cavity quantum electrodynamics [171], etc.).

For both NMR and ESR, the two-level system is generated by the Zeeman splitting of nuclear and electronic spin states, respectively, where φ_1 and φ_2 are stationary states of the system in the absence of perturbation. Once a driving field resonant with the two separated states is turned on, it acts as a perturbing Hamiltonian, $\mathscr{H}' = -\mu B_1 \cos(\omega t)$, resulting in $\mathscr{H} = \mathscr{H}_Z + \mathscr{H}'$. The static Zeeman splitting magnetic field, $\mathbf{B}_0$, gives rise to Larmor precession of frequency $\omega_0 = \frac{g\beta_B}{\hbar} \mathbf{S} \cdot \mathbf{B}_0$, whereas an analogous precession also occurs for the perturbing field, $\omega_1 = \frac{g\beta_B}{\hbar} \mathbf{S} \cdot \mathbf{B}_1$. Now, even though the system can be started in state φ_1, over time, it will evolve so that there is a finite probability of finding the system in state φ_2. See Sect. 4.6 for a discussion of this process in terms of state polarization on the Bloch sphere. In general, the time-dependence of this probability goes as $P_2(t) \propto \sin^2 \left(\frac{\Omega t}{2} \right)$, where Ω is the Rabi frequency, which is itself proportional to the off-diagonal elements of the perturbing Hamiltonian, $\Omega \propto \langle \varphi_1 | \mathscr{H}' | \varphi_2 \rangle$. By adopting a rotating frame [172] for the case at hand, the perturbing Hamiltonian becomes a constant, $\mathscr{H}' = -\mu B_1$, whose off-diagonal element computes to $\langle \varphi_1 | \mathscr{H}' | \varphi_2 \rangle = \hbar \omega_1$. This results in a Rabi frequency of $\Omega = \sqrt{\omega_1^2 + (\omega - \omega_0)^2}$, including the detuning term $(\omega - \omega_0)$ [173].

What is interesting here is that the Rabi oscillation frequency depends on the type of perturbation involved. Although the system must be resonantly driven between states φ_1 and φ_2, the presence of additional perturbations will add terms to the off-diagonal matrix elements shown above, which then directly affects the measured Rabi frequency. At this point, the observation of the Rabi frequency constitutes a spectroscopic method in its own right since it directly probes the perturbing Hamiltonian. Much work has recently been done in order to characterize this type of spectroscopy, describing the expected dependencies of Rabi frequency components from exchange[174,175] and dipolar[176,177] interactions, spin multiplicity[178], as well as simultaneously resonant pair partners[179,180]. Examples of the utility of this type of spectroscopy can be found in Chaps. 3 and 4. A more in-depth discussion concerning the practical use of this method is also given in Sects. 3.6.4 and 3.6.5.

1.2.3.2 Electron Spin Echo Envelope Modulation (ESEEM)

In order to gain the coherence lifetime, T_2, of a particular paramagnetic center, simple lineshape analysis can be sufficient if minimal environmental broadening is present [153, 181]. Very few resonant centers meet this requirement, though, and so pulsed microwave techniques have been designed for unambiguous access to this parameter. Of particular use is the Hahn echo pulse sequence [182], which probes only the phase coherence of a state while minimizing any artificial dephasing mechanisms (e.g. applied field inhomogeneities). Operation of the optically detected Hahn echo sequence in terms of pulse timing and related state polarizations on the Bloch sphere is discussed in Sects. 3.6.5 and 4.7.

In short, this pulse scheme begins with an initially polarized state,[3] say, in the singlet basis, $|Singlet\rangle$. The system is then placed into a superposition of states, $\Psi(t) = a(t)|Singlet\rangle + b(t)|Triplet\rangle$, with a $\frac{\pi}{2}$-pulse. Subsequent static dephasing is allowed to occur over some time, τ, according to the distribution of Larmor precession frequencies present. A π-pulse is then applied, effectively negating the static dephasing through time-reversal of the distributed Larmor precession. Finally, the remaining polarization, $|a(2\tau)|^2$, is monitored by projecting the state onto the observable, $PL \propto |\langle Singlet|\Psi(2\tau)\rangle|^2$. By monitoring this state polarization as a function of delay time in the pulse sequence, the characteristic time-scale for coherence, T_2, can be probed.

This simple monitoring of state decay can sometimes be complicated by the existence of perturbing local magnetic fields, from either nearby nuclear, $\mathbf{I}$, or electronic, $\mathbf{S}$, magnetic moments. The electron–nuclear dipole coupling is the most often encountered case (see Fig. 1.10). Here, the electronic moment precesses much more quickly than the nuclear moment ($\mu_e \gg \mu_n$), and so changes made to its spin orientation with the microwave pulse occur on a nonadiabatic time-scale for the nuclear moment. The nuclear moment then precesses around its new local field, determined by the new orientation of the much larger electronic moment. This

---

[3]In general, "state polarization" refers to the magnitude of content for a particular eigenstate of the system. In ESR, this corresponds to the actual polarization of magnetic moments, whereas in ODMR, it is the amount of singlet or triplet content within the system.

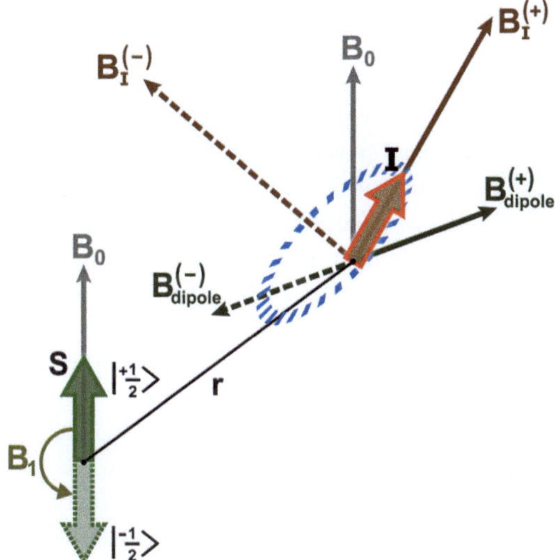

Fig. 1.10 A classical model for electron–nuclear ESEEM. In magnetic dipole–dipole hyperfine coupling, an $\mathbf{S} = \tfrac{1}{2}$ electron interacts with a nuclear moment, $\mathbf{I}$, a distance $\mathbf{r}$ away. Since the local field generated by the electron ($\mathbf{B}_{dipole}^{(+)}$) is much larger than that generated by the nucleus, the nuclear moment experiences an effective field, $\mathbf{B}_I^{(+)}$, when the electron is in a spin-up state. Upon inducing a spin flip with an RF field, $\mathbf{B}_1$, the effective field experienced at the nucleus suddenly changes to $\mathbf{B}_I^{(-)}$. The weak nuclear moment then begins to Larmor precess about the new field position. The precession of this small nuclear moment acts as a slow periodic perturbation to the electron moment. Since the Hahn echo decay experiment is designed to be sensitive to changes in state polarization, this periodic perturbation expresses itself as a modulation in the decay of electronic state polarization

nuclear precession acts as a small perturbation on the electronic moment, causing it to be periodically out of phase with the initially established polarization during the pulse sequence. In the end, the envelope of decay for state polarization, as measured by spin echo amplitude, is observed to have some periodic "loss" in coherence. The effect is then known as electron spin echo envelope modulation, or ESEEM.

Since the frequency of this periodic dephasing is directly determined by the magnetic moment of the nucleus, measuring ESEEM in the Hahn echo decay of a system can be used as an environmental probe. In fact, ESEEM is regularly used as a spectroscopic technique in this way, through traditional polarization [183–186], as well as optical [159, 187, 188] and electrical [189] detection schemes. An excellent resource for the quantum mechanical treatment of this effect under various environmental conditions can be found in Ref. [190].

1.2.3.3   Decoherence in Nanocrystals

Spin dynamics in semiconductor nanocrystals are normally quite different than
for the same states within bulk versions of these materials. In bulk semiconductors, spin states normally have a very short lifetime (100 fs [191, 192] to a few
nanoseconds [192]), due to the domination of spin-orbit coupling as a perturbation.
In nanoscale systems, though, quantum confinement leads to the discretization of
energy bands, as discussed in Sect. 1.1.1. Since the separation of energy levels
becomes comparatively large to the bulk case, spin-orbit interaction becomes weak
as a perturbation (see Sect. 1.1.2.1). Room temperature spin dynamics are still
limited to short time-scales [65], though, due to phonon-mediated spin flips. These
spin decay mechanisms typically involve a 2-phonon transition from one spin
state to the other, mediated by either real phonons (Orbach process) or virtual
phonons (Raman process) [193]. At low temperatures ($T < 10$ K), phonon modes
are largely frozen out, making 2-phonon processes unlikely. A so-called "direct"
phonon process between spin states is also possible, but meeting the condition
of energy matching with single phonons is not probable. Instead, the slowly
fluctuating hyperfine interactions between band edge carriers and the surrounding
nuclear moments have been found to be the dominant dephasing mechanism for
semiconductor quantum dots [194]. This longer time-scale allows for spin coherence
to routinely persist over a few nanoseconds for pristine nanocrystals of several
material types [153, 195, 196].

A distinction must be made between the spin dynamics observed for band edge
carrier states and of the trapped carrier states discussed in Chaps. 3 and 4. Much
of the literature concerning the coherence of electronic states in nanocrystals has
used optical probes in their characterizations [195–197], limiting their studies to
excitonic and band gap states. Carriers which occupy the electronic states of the
nanocrystal are inherently delocalized across it, which then leads to the hyperfine
mediated spin dephasing just described. For the highly localized states which are
considered in this dissertation, though, such an interaction with many nuclear sites
is absent. As expected, the coherence lifetime for these states is correspondingly
increased, surpassing those which have been reported to date for this material system
[153, 198]. Since this work is an early investigation into the nature of these types of
carrier states, there presently exists no information on the relevant perturbations
leading to spin dephasing and decay.

References

1. Feynman, R. P.: There's plenty of room at the bottom. J. Microelectromech. Syst. **1**(1), 60–66
   (1992)
2. Binning, G., Rohrer, H.: Scanning tunneling microscopy. IBM J. Res. Dev. **30**(4), 355–369
   (1986)
3. Taniguchi, N.: Current status in, and future trends of, ultraprecision machining and ultrafine
   materials processing. CIRP Ann. Manuf. Technol. **32**(2), 573–582 (1983)

4. Martin, P.J., Macleod, H.A., Netterfield, R.P., Pacey, C.G., Sainty, W.G.: Ion-beam-assisted deposition of thin films. Appl. Opt. **22**(1), 178–184 (1983)
5. Freestone, I., Meeks, N., Sax, M., Higgitt, C.: The Lycurgus Cup – a Roman nanotechnology. Gold Bull. **40**(4), 270–277 (2007)
6. Barber, D.J., Freestone, I.C.: An investigation of the origin of the colour of the Lycurgus Cup by analytical transmission electron microscopy. Archaeometry **32**(1), 33–45 (1990)
7. Granqvist, C., Buhrman, R., Wyns, J., Sievers, A.: Far-infrared absorption in ultrafine Al particles. Phys. Rev. Lett. **37**(10), 625–629 (1976)
8. Ekimov, A., Onushchenko, A.: Quantum size effect in three-dimensional microscopic semiconductor crystals. JETP Lett. **34**(6), 345–349 (1981)
9. Kouwenhoven, L., Johnson, A., van der Vaart, N., Harmans, C., Foxon, C.: Quantized current in a quantum-dot turnstile using oscillating tunnel barriers. Phys. Rev. Lett. **67**(12), 1626–1629 (1991)
10. Patton, B., Langbein, W., Woggon, U.: Trion, biexciton, and exciton dynamics in single self-assembled CdSe quantum dots. Phys. Rev. B **68**(12), 125316 (2003)
11. Cölfen, H., Antonietti, M.: Mesocrystals: inorganic superstructures made by highly parallel crystallization and controlled alignment. Angew. Chem. Int. Ed. (English) **44**(35), 5576–5591 (2005)
12. Gutiérrez, M., Henglein, A.: Photochemistry of colloidal metal Sulfides. 4. Cathodic dissolution of CdS and excess Cd2+ reduction. Ber. Bunsenges. Phys. Chem. **87**(6), 474–478 (1983)
13. Murray, C.B., Norris, D.J., Bawendi, M.G.: Synthesis and characterization of nearly monodisperse CdE (E = S, Se, Te) semiconductor nanocrystallites. J. Am. Chem. Soc. **115**(19), 8706–8715 (1993)
14. Mićić, O., Nozik, A.: Synthesis and characterization of binary and ternary III–V quantum dots. J. Lumin. **70**(1–6), 95–107 (1996)
15. Li, Y.-D., Liao, H.-W., Ding, Y., Qian, Y.-T., Yang, L., Zhou, G.-E.: Nonaqueous synthesis of CdS nanorod semiconductor. Chem. Mater. **10**(9), 2301–2303 (1998)
16. Zhang, H., Hyun, B.-R., Wise, F.W., Robinson, R.D.: A generic method for rational scalable synthesis of monodisperse metal sulfide nanocrystals. Nano Lett. **12**, 5856–5860 (2012)
17. Sargent, E.: Colloidal quantum dot solar cells. Nat. Photon. **6**, 133–135 (2012)
18. Wood, V., Panzer, M.J., Chen, J., Bradley, M.S., Halpert, J.E., Bawendi, M.G., Bulović, V: Inkjet-printed quantum dot-polymer composites for full-color AC-driven displays. Adv. Mater. **21**(21), 2151–2155 (2009)
19. Kim, T.-H., Cho, K.-S., Lee, E.K., Lee, S.J., Chae, J., Kim, J.W., Kim, D.H., Kwon, J.-Y., Amaratunga, G., Lee, S.Y., Choi, B.L., Kuk, Y., Kim, J.M., Kim, K.: Full-colour quantum dot displays fabricated by transfer printing. Nat. Photon. **5**(3), 176–182 (2011)
20. Kramer, I.J., Sargent, E.H.: Colloidal quantum dot photovoltaics: a path forward. ACS Nano **5**(11), 8506–8514 (2011)
21. Semonin, O.E., Luther, J.M., Choi, S., Chen, H.-Y., Gao, J., Nozik, A.J., Beard, M.C.: Peak external photocurrent quantum efficiency exceeding 100% via MEG in a quantum dot solar cell. Science **334**(6062), 1530–1533 (2011)
22. Tang, J., Kemp, K.W., Hoogland, S., Jeong, K.S., Liu, H., Levina, L., Furukawa, M., Wang, X., Debnath, R., Cha, D., Chou, K.W., Fischer, A., Amassian, A., Asbury, J.B., Sargent, E.H.: Colloidal-quantum-dot photovoltaics using atomic-ligand passivation. Nat. Mater. **10**(10), 765–771 (2011)
23. Ip, A.H., Thon, S.M., Hoogland, S., Voznyy, O., Zhitomirsky, D., Debnath, R., Levina, L., Rollny, L.R., Carey, G.H., Fischer, A., Kemp, K.W., Kramer, I.J., Ning, Z., Labelle, A.J., Chou, K.W., Amassian, A., Sargent, E.H.: Hybrid passivated colloidal quantum dot solids. Nat. Nanotechnol. **7**(9), 577–582 (2012)
24. Zhitomirsky, D., Kramer, I.J., Labelle, A.J., Fischer, A., Debnath, R., Pan, J., Bakr, O.M., Sargent, E.H.: Colloidal quantum dot photovoltaics: the effect of polydispersity. Nano Lett. **12**(2), 1007–1012 (2012)

25. Dang, C., Lee, J., Breen, C., Steckel, J.S., Coe-Sullivan, S., Nurmikko, A.: Red, green and blue lasing enabled by single-exciton gain in colloidal quantum dot films. Nat. Nanotechnol. **7**(5), 335–339 (2012)
26. Seydel, C.: Quantum dots get wet. Science **300**(5616), 80–81 (2003)
27. Rosen, S., Schwartz, O., Oron, D.: Transient fluorescence of the off state in blinking CdSe/CdS/ZnS semiconductor nanocrystals is not governed by Auger recombination. Phys. Rev. Lett. **104**(15), 157404 (2010)
28. Zhao, J., Nair, G., Fisher, B.R., Bawendi, M.G.: Challenge to the charging model of semiconductor-nanocrystal fluorescence intermittency from off-state quantum yields and multiexciton blinking. Phys. Rev. Lett. **104**(15), 157403 (2010)
29. Galland, C., Ghosh, Y., Steinbrück, A., Sykora, M., Hollingsworth, J.A., Klimov, V.I., Htoon, H.: Two types of luminescence blinking revealed by spectroelectrochemistry of single quantum dots. Nature **479**(7372), 203–207 (2011)
30. Jones, M., Lo, S.S., Scholes, G.D.: Quantitative modeling of the role of surface traps in CdSe/CdS/ZnS nanocrystal photoluminescence decay dynamics. Proc. Natl. Acad. Sci. **106**(9), 3011–3016 (2009)
31. Liu, S., Borys, N., Huang, J., Talapin, D., Lupton, J.: Exciton storage in CdSe/CdS tetrapod semiconductor nanocrystals: electric field effects on exciton and multiexciton states. Phys. Rev. B **86**(4), 045303 (2012)
32. Trinh, M.T., Limpens, R., de Boer, W.D.A.M., Schins, J.M., Siebbeles, L.D.A., Gregorkiewicz, T.: Direct generation of multiple excitons in adjacent silicon nanocrystals revealed by induced absorption. Nat. Photon. **6**(5), 316–321 (2012)
33. Efros, A.L., Rosen, M.: Random telegraph signal in the photoluminescence intensity of a single quantum dot. Phys. Rev. Lett. **78**(6), 1110–1113 (1997)
34. van Driel, A.F., Nikolaev, I.S., Vergeer, P., Lodahl, P., Vanmaekelbergh, D., Vos, W.L.: Statistical analysis of time-resolved emission from ensembles of semiconductor quantum dots: interpretation of exponential decay models. Phys. Rev. B **75**(3), 035329 (2007)
35. Cooney, R.R., Sewall, S.L., Anderson, K.E.H., Dias, E.A., Kambhampati, P.: Breaking the phonon bottleneck for holes in semiconductor quantum dots. Phys. Rev. Lett. **98**(17), 177403 (2007)
36. Pandey, A.K., Guyot-Sionnest, P.: Slow electron cooling in colloidal quantum dots. Science **322**, 929–932 (2008)
37. Kittel, C.: Introduction to Solid State Physics, 8th edn., pp. 161–183. Wiley, Hoboken, NJ, USA (2005)
38. Scholes, G.D., Rumbles, G.: Excitons in nanoscale systems. Nat. Mater. **5**(9), 683–696 (2006)
39. Kittel, C.: Introduction to Solid State Physics, 441, 8th edn. Wiley, Hoboken, NJ, USA (2005)
40. Brus, L.E.: Electron–electron and electron-hole interactions in small semiconductor crystallites: the size dependence of the lowest excited electronic state. J. Chem. Phys. **80**(9), 4403–4409 (1984)
41. Brus, L.: Electronic wave functions in semiconductor clusters: experiment and theory. J. Phys. Chem. **90**(12), 2555–2560 (1986)
42. Beaulac, R., Schneider, L., Archer, P.I., Bacher, G., Gamelin, D.R.: Light-induced spontaneous magnetization in doped colloidal quantum dots. Science **325**(5943), 973–976 (2009)
43. Ochsenbein, S.T., Feng, Y., Whitaker, K.M., Badaeva, E., Liu, W.K., Li, X., Gamelin, D.R.: Charge-controlled magnetism in colloidal doped semiconductor nanocrystals. Nat. Nanotechnol. **4**(10), 681–687 (2009)
44. Smith, A.M., Mohs, A.M., Nie, S.: Tuning the optical and electronic properties of colloidal nanocrystals by lattice strain. Nat. Nanotechnol. **4**(1), 56–63 (2009)
45. Park, K., Deutsch, Z., Li, J.J., Oron, D., Weiss, S.: Single molecule quantum-confined Stark effect measurements of semiconductor nanoparticles at room temperature. ACS Nano **6**, 10013–10023 (2012)
46. Brovelli, S., Schaller, R.D., Crooker, S.A., García-Santamaría, F., Chen, Y., Viswanatha, R., Hollingsworth, J.A., Htoon, H., Klimov, V.I.: Nano-engineered electron-hole exchange interaction controls exciton dynamics in core-shell semiconductor nanocrystals. Nat. Com-

mun. **2**, 280 (2011)

47. Müller, J., Lupton, J.M., Lagoudakis, P.G., Schindler, F., Koeppe, R., Rogach, A.L., Feldmann, J., Talapin, D.V., Weller, H.: Wave function engineering in elongated semiconductor nanocrystals with heterogeneous carrier confinement. Nano Lett. **5**(10), 2044–2049 (2005)

48. Borys, N.J., Walter, M.J., Huang, J., Talapin, D.V., Lupton, J.M.: The role of particle morphology in interfacial energy transfer in CdSe/CdS heterostructure nanocrystals. Science **330**, 1371–1374 (2010)

49. Jun, Y.-W., Lee, J.-H., Choi, J.-S., Cheon, J.: Symmetry-controlled colloidal nanocrystals: nonhydrolytic chemical synthesis and shape determining parameters. J. Phys. Chem. B **109**(31), 14795–14806 (2005)

50. Talapin, D.V., Nelson, J.H., Shevchenko, E.V., Aloni, S., Sadtler, B., Alivisatos, A.P.: Seeded growth of highly luminescent CdSe/CdS nanoheterostructures with rod and tetrapod morphologies. Nano Lett. **7**(10), 2951–2959 (2007)

51. Srivastava, S., Santos, A., Critchley, K., Kim, K.-S., Podsiadlo, P., Sun, K., Lee, J., Xu, C., Lilly, G.D., Glotzer, S.C., Kotov, N.A.: Light-controlled self-assembly of semiconductor nanoparticles into twisted ribbons. Science **327**(5971), 1355–1359 (2010)

52. Milliron, D.J., Hughes, S.M., Cui, Y., Manna, L., Li, J., Wang, L.-W., Alivisatos, A.P.: Colloidal nanocrystal heterostructures with linear and branched topology. Nature **430**(6996), 190–195 (2004)

53. Bawendi, M., Wilson, W., Rothberg, L., Carroll, P., Jedju, T., Steigerwald, M., Brus, L.: Electronic structure and photoexcited-carrier dynamics in nanometer-size CdSe clusters. Phys. Rev. Lett. **65**(13), 1623–1626 (1990)

54. Bawendi, M.G., Carroll, P.J., Wilson, W.L., Brus, L.E.: Luminescence properties of CdSe quantum crystallites: resonance between interior and surface localized states. J. Chem. Phys. **96**(2), 946–954 (1992)

55. Henry, C., Nassau, K.: Lifetimes of bound excitons in CdS. Phys. Rev. B **1**(4), 1628–1634 (1970)

56. Efros, A., Rosen, M., Kuno, M., Nirmal, M., Norris, D., Bawendi, M.: Band-edge exciton in quantum dots of semiconductors with a degenerate valence band: dark and bright exciton states. Phys. Rev. B **54**(7), 4843–4856 (1996)

57. Nirmal, M., Norris, D., Kuno, M., Bawendi, M., Efros, A., Rosen, M.: Observation of the "dark exciton" in CdSe quantum dots. Phys. Rev. Lett. **75**(20), 3728–3731 (1995)

58. Efros, A., Rosen, M.: Quantum size level structure of narrow-gap semiconductor nanocrystals: effect of band coupling. Phys. Rev. B **58**(11), 7120–7135 (1998)

59. Rodina, A., Efros, A., Alekseev, A.: Effect of the surface on the electron quantum size levels and electron g-factor in spherical semiconductor nanocrystals. Phys. Rev. B **67**(15), 155312 (2003)

60. Sapra, S., Sarma, D.: Evolution of the electronic structure with size in II–VI semiconductor nanocrystals. Phys. Rev. B **69**(12), 125304 (2004)

61. Planelles, J., Rajadell, F., Climente, J.I.: Hole band mixing in CdS and CdSe quantum dots and quantum rods. J. Phys. Chem. C **114**(18), 8337–8342 (2010)

62. Horodyská, P., Němec, P., Sprinzl, D., Malý, P., Gladilin, V.N., Devreese, J.T.: Exciton spin dynamics in spherical CdS quantum dots. Phys. Rev. B **81**(4), 045301 (2010)

63. Norris, D.J.: Electronic structure in semiconductor nanocrystals: optical experiment. In: Klimov, V.I. (ed.) Nanocrystal Quantum Dots, chapter 2, 2nd edn., pp. 63–96. CRC, Boca Raton, FL, USA (2010)

64. Richard, T., Lefebvre, P., Mathieu, H., Allègre, J.: Effects of finite spin-orbit splitting on optical properties of spherical semiconductor quantum dots. Phys. Rev. B **53**(11), 7287–7298 (1996)

65. Huxter, V.M., Kim, J., Lo, S.S., Lee, A., Nair, P.S., Scholes, G.D.: Spin relaxation in zinc blende and wurtzite CdSe quantum dots. Chem. Phys. Lett. **491**(4–6), 187–192 (2010)

66. Nirmal, M., Dabbousi, B.O., Bawendi, M.G., Macklin, J.J., Trautman, J.K., Harris, T.D., Brus, L.E.: Fluorescence intermittency in single cadmium selenide nanocrystals. Nature **383**(6603),

802–804 (1996)

67. Shimizu, K., Neuhauser, R., Leatherdale, C., Empedocles, S., Woo, W., Bawendi, M.: Blinking statistics in single semiconductor nanocrystal quantum dots. Phys. Rev. B **63**(20), 205316 (2001)

68. Cook, R., Kimble, H.: Possibility of direct observation of quantum jumps. Phys. Rev. Lett. **54**(10), 1023–1026 (1985)

69. Vanden Bout, D.A., Yip, W.-T., Hu, D., Fu, D.-K., Swager, T.M., Barbara, P.F.: Discrete intensity jumps and intramolecular electronic energy transfer in the spectroscopy of single conjugated polymer molecules. Science **277**(5329), 1074–1077 (1997)

70. Jones, M., Scholes, G.D.: On the use of time-resolved photoluminescence as a probe of nanocrystal photoexcitation dynamics. J. Mater. Chem. **20**(18), 3533–3538 (2010)

71. Fisher, B.R., Eisler, H.-J., Stott, N.E., Bawendi, M.G.: Emission intensity dependence and single-exponential behavior in single colloidal quantum dot fluorescence lifetimes. J. Phys. Chem. B **108**(1), 143–148 (2004)

72. Knappenberger, K.L., Wong, D.B., Romanyuk, Y.E., Leone, S.R.: Excitation wavelength dependence of fluorescence intermittency in CdSe/ZnS core/shell quantum dots. Nano Lett. **7**(12), 3869–3874 (2007)

73. Kim, Y., Song, N.W., Yu, H., Moon, D.W., Lim, S.J., Kim, W., Yoon, H.-J., Shin, S.K.: Ligand-dependent blinking of zinc-blende CdSe/ZnS core/shell nanocrystals. Phys. Chem. Chem. Phys. **11**, 3497–3502 (2009)

74. Ishizumi, A., Kanemitsu, Y.: Blinking behavior of surface-defect and impurity luminescence in nondoped and Mn 2+ -doped CdS nanocrystals. J. Phys. Soc. Jpn. **78**(8), 083705 (2009)

75. Frantsuzov, P., Kuno, M., Jankó, B., Marcus, R.A.: Universal emission intermittency in quantum dots, nanorods and nanowires. Nat. Phys. **4**(5), 519–522 (2008)

76. Peterson, J.J., Nesbitt, D.J.: Modified power law behavior in quantum dot blinking: a novel role for biexcitons and auger ionization. Nano Lett. **9**(1), 338–345 (2009)

77. Spinicelli, P., Buil, S., Quelin, X., Mahler, B., Dubertret, B., Hermier, J.-P.: Bright and grey states in CdSe-CdS nanocrystals exhibiting strongly reduced blinking. Phys. Rev. Lett. **102**(13), 136801 (2009)

78. Zhang, K., Chang, H., Fu, A., Alivisatos, A.P., Yang, H.: Continuous distribution of emission states from single CdSe/ZnS quantum dots. Nano Lett. **6**(4), 843–847 (2006)

79. Gómez-Campos, F.M., Califano, M.: Hole surface trapping in CdSe nanocrystals: dynamics, rate fluctuations, and implications for blinking. Nano Lett. **12**(9), 4508–4517 (2012)

80. Wang, X., Ren, X., Kahen, K., Hahn, M.A., Rajeswaran, M., Maccagnano-Zacher, S., Silcox, J., Cragg, G.E., Efros, A.L., Krauss, T.D.: Non-blinking semiconductor nanocrystals. Nature **459**(7247), 686–689 (2009)

81. Chon, B., Lim, S.J., Kim, W., Seo, J., Kang, H., Joo, T., Hwang, J., Shin, S.K.: Shell and ligand-dependent blinking of CdSe-based core/shell nanocrystals. Phys. Chem. Chem. Phys. **12**(32), 9312–9319 (2010)

82. Issac, A., Krasselt, C., Cichos, F., von Borczyskowski, C.: Influence of the dielectric environment on the photoluminescence intermittency of CdSe quantum dots. Chemphyschem **13**(13), 3223–3230 (2012)

83. Schroeter, D., Griffiths, D., Sercel, P.: Defect-assisted relaxation in quantum dots at low temperature. Phys. Rev. B **54**(3), 1486–1489 (1996)

84. Schreuder, M.A., Xiao, K., Ivanov, I.N., Weiss, S.M., Rosenthal, S.J.: White light-emitting diodes based on ultrasmall CdSe nanocrystal electroluminescence. Nano. Lett **10**(2), 573–576 (2010)

85. Chan, Y., Steckel, J.S., Snee, P.T., Caruge, J.-M., Hodgkiss, J.M., Nocera, D.G., Bawendi, M.G.: Blue semiconductor nanocrystal laser. Appl. Phys. Lett. **86**(7), 073102 (2005)

86. Voznyy, O.: Mobile surface traps in CdSe nanocrystals with carboxylic acid ligands. J. Phys. Chem. C **115**(32), 15927–15932 (2011)

87. Fischer, S.A., Crotty, A.M., Kilina, S.V., Ivanov, S.A., Tretiak, S.: Passivating ligand and solvent contributions to the electronic properties of semiconductor nanocrystals. Nanoscale **4**(3), 904–914 (2012)

88. Henrich, V.: The nature of transition-metal-oxide surfaces. Prog. Surf. Sci. **14**(2), 175–199 (1983)
89. Wei, H.H.-Y., Evans, C.M., Swartz, B.D., Neukirch, A.J., Young, J., Prezhdo, O.V., Krauss, T.D.: Colloidal semiconductor quantum dots with tunable surface composition. Nano Lett. **12**(9), 4465–4471 (2012)
90. Morfa, A.J., Gibson, B.C., Karg, M., Karle, T.J., Greentree, A.D., Mulvaney, P., Tomljenovic-Hanic, S.: Single-photon emission and quantum characterization of zinc oxide defects. Nano Lett. **12**(2), 949–954 (2012)
91. Klimov, V., Bolivar, P.H., Kurz, H.: Ultrafast carrier dynamics in semiconductor quantum dots. Phys. Rev. B **53**(3), 1463–1467 (1996)
92. Kim, D., Mishima, T., Tomihira, K., Nakayama, M.: Temperature dependence of photoluminescence dynamics in colloidal CdS quantum dots. J. Phys. Chem. C **112**(29), 10668–10673 (2008)
93. Vuylsteke, A., Sihvonen, Y.: Sulfur vacancy mechanism in pure CdS. Phys. Rev. **113**(1), 40–42 (1959)
94. Edgar, A., Pörsch, J.: Optically detected magnetic resonance from a complex donor in CdS. Solid State Commun. **44**(5), 741–743 (1982)
95. Babentsov, V., Riegler, J., Schneider, J., Ehlert, O., Nann, T., Fiederle, M.: Deep level defect luminescence in cadmium selenide nano-crystals films. J. Cryst. Growth **280**(3-4), 502–508 (2005)
96. Chestnoy, N., Harris, T., Hull, R., Brus, L.: Luminescence and photophysics of cadmium sulfide semiconductor clusters: the nature of the emitting electronic state. J. Phys. Chem. **90**(15), 3393–3399 (1986)
97. Babentsov, V., Sizov, F.: Defects in quantum dots of IIB–VI semiconductors. Opto-Electron. Rev. **16**(3), 208–225 (2008)
98. Kuçur, E., Bücking, W., Giernoth, R., Nann, T.: Determination of defect states in semiconductor nanocrystals by cyclic voltammetry. J. Phys. Chem. B **109**(43), 20355–20360 (2005)
99. Brovelli, S., Galland, C., Viswanatha, R., Klimov, V.I.: Tuning radiative recombination in Cu-doped nanocrystals via electrochemical control of surface trapping. Nano. Lett. **12**(8), 4372–4379 (2012)
100. Alperson, B., Rubinstein, I., Hodes, G.: Identification of surface states on individual CdSe quantum dots by room-temperature conductance spectroscopy. Phys. Rev. B **63**(8), 081303(R) (2001)
101. Magno, R., Bennett, B.R., Glaser, E.R.: Deep level transient capacitance measurements of GaSb self-assembled quantum dots. J. Appl. Phys. **88**(10), 5843–5849 (2000)
102. Pincherle, L.: Energy levels of F-centres. Proc. Phys. Soc. A **64**(7), 648–657 (1951)
103. Li, D., Zhang, J., Zhang, Q., Xiong, Q.: Electric-field-dependent photoconductivity in CdS nanowires and nanobelts: exciton ionization, Franz-Keldysh, and Stark effects. Nano Lett. **12**(6), 2993–2999 (2012)
104. Brunwin, R.F., Cavenett, B.C., Davies, J.J., Nicholls, J.E.: Optically detected donor magnetic resonance in CdS. Solid State Commun. **18**(9–10), 1283–1285 (1976)
105. Lifshitz, E., Dag, I., Litvitn, I.D., Hodes, G.: Optically detected magnetic resonance study of electron/hole traps on CdSe quantum dot surfaces. J. Phys. Chem. B **102**(46), 9245–9250 (1998)
106. Litvin, I.D., Porteanu, H., Lifshitz, E., Lipovskii, A.A.: Optically detected magnetic resonance studies of CdS nanoparticles grown in phosphate glass. J. Cryst. Growth **198–199**(Part 1), 313–315 (1999)
107. Lifshitz, E., Porteanu, H., Glozman, A., Weller, H., Pflughoefft, M., Echymuller, A.: Optically detected magnetic resonance study of CdS/HgS/CdS quantum dot quantum wells. J. Phys. Chem. B **103**(33), 6870–6875 (1999)
108. Lifshitz, E., Glozman, A., Litvin, I.D., Porteanu, H.: Optically detected magnetic resonance studies of the surface/interface properties of II–VI semiconductor quantum dots. J. Phys. Chem. B **104**(45), 10449–10461 (2000)

109. Glozman, A., Lifshitz, E.: Optically detected spin and spin-orbital resonance studies of CdSe/CdS core-shell nanocrystals. Mater. Sci. Eng. C **15**(1–2), 17–19 (2001)
110. Lifshitz, E., Glozman, A.: Optically detected spin and orbit resonance of semiconductor quantum dots. Phys. Status Solidi B **224**(2), 541–544 (2001)
111. Langof, L., Ehrenfreund, E., Lifshitz, E., Micic, O.I., Nozik, A.J.: Continuous-wave and time-resolved optically detected magnetic resonance studies of nonetched/etched InP nanocrystals. J. Phys. Chem. B **106**(7), 1606–1612 (2002)
112. Gokhberg, K., Glozman, A., Lifshitz, E., Maniv, T., Schlamp, M.C., Alivisatos, P.: Electron (hole) paramagnetic resonance of spherical CdSe nanocrystals. J. Chem. Phys. **117**(6), 2909–2913 (2002)
113. Lifshitz, E., Fradkin, L., Glozman, A., Langof, L.: Optically detected magnetic resonance studies of colloidal semiconductor nanocrystals. Annu. Rev. Chem. **55**, 509–557 (2004)
114. Fradkin, L., Langof, L., Lifshitz, E., Gaponik, N., Rogach, A., Eychmüller, A., Weller, H., Mićić, O.I., Nozik, A.J.: A direct measurement of g-factors in II–VI and III–V core-shell nanocrystals. Phys. E **26**(1-4), 9–13 (2005)
115. Zeeman, P.: The effect of magnetisation on the nature of light emitted by a substance. Nature **55**(1424), 347–347 (1897)
116. Hale, G.E.: On the probable existence of a magnetic field in sun-spots. Astrophys. J. **28**, 315 (1908)
117. Hale, G.E., Ellerman, F., Nicholson, S.B., Joy, A.H.: The magnetic polarity of sun-spots. Astrophys. J. **49**, 153 (1919)
118. Al'tshuler, S., Zavoisky, E.K., Kozyrev, B.: New method to study paramagnetic absorption. Zhurn. Eksperiment. Teoret. Fiziki **14**(10–11), 407–409 (1944)
119. Cummerow, R., Halliday, D.: Paramagnetic losses in two manganous salts. Phys. Rev. **70**(5–6), 433–433 (1946)
120. Cummerow, R., Halliday, D., Moore, G.: Paramagnetic resonance absorption in salts of the iron group. Phys. Rev. **72**(12), 1233–1240 (1947)
121. Holden, A., Kittel, C., Merritt, F., Yager, W.: Microwave resonance absorption in a paramagnetic organic compound. Phys. Rev. **75**(10), 1614–1614 (1949)
122. Hutchison, C.A.: Paramagnetic resonance absorption in crystals colored by irradiation. Phys. Rev. **75**(11), 1769–1770 (1949)
123. Abragam, A.: Paramagnetic resonance and hyperfine structure in the iron transition group. Phys. Rev. **79**(3), 534–534 (1950)
124. Elliott, R.: Crystal field theory in the rare earths. Rev. Mod. Phys. **25**(1), 167–169 (1953)
125. Loss, D., Burkard, G., DiVincenzo, D.P.: Electron spins in quantum dots as quantum bits. J. Nanopart. Res. **2**(4), 401–411 (2000)
126. Neumann, P., Kolesov, R., Naydenov, B., Beck, J., Rempp, F., Steiner, M., Jacques, V., Balasubramanian, G., Markham, M.L., Twitchen, D.J., Pezzagna, S., Meijer, J., Twamley, J., Jelezko, F., Wrachtrup, J.: Quantum register based on coupled electron spins in a room-temperature solid. Nat. Phys. **6**(4), 249–253 (2010)
127. Pla, J.J., Tan, K.Y., Dehollain, J.P., Lim, W.H., Morton, J.J.L., Jamieson, D.N., Dzurak, A.S., Morello, A.: A single-atom electron spin qubit in silicon. Nature **489**(7417), 541–545 (2012)
128. Kane, B.: A silicon-based nuclear spin quantum computer. Nature **393**, 133–137 (1998)
129. Ladd, T.D., Jelezko, F., Laflamme, R., Nakamura, Y., Monroe, C., O'Brien, J.L.: Quantum computers. Nature **464**(7285), 45–53 (2010)
130. McCamey, D.R., Van Tol, J., Morley, G.W., Boehme, C.: Electronic spin storage in an electrically readable nuclear spin memory with a lifetime >100 seconds. Science **330**(6011), 1652–1656 (2010)
131. Wrachtrup, J., von Borczyskowski, C., Bernard, J., Orrit, M., Brown, R.: Optical detection of magnetic resonance in a single molecule. Nature **363**, 244–245 (1993)
132. Shkrob, I.A., Trifunac, A.D.: Magnetic resonance and spin dynamics in radical ion pairs: pulsed time-resolved fluorescence detected magnetic resonance. J. Chem. Phys. **103**(2), 551–561 (1995)

133. Breiland, W.G.: Coherence in multilevel systems. I. Coherence in excited states and its application to optically detected magnetic resonance in phosphorescent triplet states. J. Chem. Phys. **62**(9), 3458–3475 (1975)

134. Chaudhuri, D., Wettach, H., van Schooten, K.J., Liu, S., Sigmund, E., Höger, S., Lupton, J.M.: Tuning the singlet-triplet gap in metal-free phosphorescent π–conjugated polymers. Angew. Chem. Int. Ed. (English) **49**(42), 7714–7717 (2010)

135. Chaudhuri, D., Li, D., Sigmund, E., Wettach, H., Höger, S., Lupton, J.M.: Plasmonic surface enhancement of dual fluorescence and phosphorescence emission from organic semiconductors: effect of exchange gap and spin-orbit coupling. Chem. Commun. **48**(53), 6675–6677 (2012)

136. Boehme, C., Lips, K.: Theory of time-domain measurement of spin-dependent recombination with pulsed electrically detected magnetic resonance. Phys. Rev. B **68**(24), 245105 (2003)

137. Brodsky, S., Franke, V., Hiller, J., McCartor, G., Paston, S., Prokhvatilov, E.: A nonperturbative calculation of the electron's magnetic moment. Nucl. Phys. B **703**(1-2), 333–362 (2004)

138. Odom, B., Hanneke, D., D'Urso, B., Gabrielse, G.: New measurement of the electron magnetic moment using a one-electron quantum cyclotron. Phys. Rev. Lett. **97**(3), 030801 (2006)

139. Gabrielse, G., Hanneke, D., Kinoshita, T., Nio, M., Odom, B.: New determination of the fine structure constant from the electron g value and QED. Phys. Rev. Lett. **97**(3), 10–13 (2006)

140. Atherton, N.M.: Principles of Electron Spin Resonance, 44. Ellis Horwood Limited, London, Chichester, UK (1993)

141. Weil, J.A., Bolton, J.R.: Electron Paramagnetic Resonance: Elementary Theory and Practical Applications, 2nd edn., pp. 105–110. Wiley, Hoboken, NJ, USA (2007)

142. Bennati, M., Murphy, D.M.: Electron paramagnetic resonance spectra in the solid state. In: Brustolon, M., Giamello, E. (eds.) Electron Paramagnetic Resonance: A Practitioner's Toolkit, chapter 6, pp. 207–209. Wiley, Hoboken, NJ, USA (2009)

143. Nilsson, H.A., Caroff, P., Thelander, C., Larsson, M., Wagner, J.B., Wernersson, L.-E., Samuelson, L., Xu, H.Q.: Giant, level-dependent g factors in InSb nanowire quantum dots. Nano Lett. **9**(9), 3151–3156 (2009)

144. Kohmoto, T., Fukuda, Y., Hashi, T.: Optically detected low-field ESR in the metastable state of $Tm^{2+}:SrF_2$. II. Determination of the sign of the g factor. Phys. Rev. B **34**(9), 6094–6098 (1986)

145. Vana, N.J., Unfried, E.: A precise g-factor standard for ESR measurements in a wide temperature range. J. Magn. Reson. **6**(4), 655–659 (1972)

146. Frait, Z., Gemperle, R.: Precisiong-factor measurements of electron paramagnetic resonance standards. Czech. J. Phys. **27**(1), 99–112 (1977)

147. Paik, S.-Y., Lee, S.-Y., Baker, W.J., McCamey, D. R., Boehme, C.: T1 and T2 spin relaxation time limitations of phosphorous donor electrons near crystalline silicon to silicon dioxide interface defects. Phys. Rev. B **81**(7), 075214 (2010)

148. Feher, G.: Electron spin resonance experiments on donors in silicon. I. Electronic structure of donors by the electron nuclear double resonance technique. Phys. Rev. **114**(5), 1219 (1959)

149. Cullis, P., Marko, J.: Electron paramagnetic resonance properties of n-type silicon in the intermediate impurity-concentration range. Phys. Rev. B **11**(11), 4184 (1975)

150. Baker, W.J., Ambal, K., Waters, D.P., Baarda, R., Morishita, H., van Schooten, K., McCamey, D.R., Lupton, J.M., Boehme, C.: Robust absolute magnetometry with organic thin-film devices. Nat. Commun. **3**(898) (2012). DOI 10.1038/ncomms1895

151. Schoenfeld, R., Harneit, W.: Real time magnetic field sensing and imaging using a single spin in diamond. Phys. Rev. Lett. **106**(3), 030802 (2011)

152. Weil, J.A., Bolton, J.R.: Electron Paramagnetic Resonance: Elementary Theory and Practical Applications, 2nd edn. Wiley, Hoboken, NJ, USA (2007)

153. Whitaker, K.M., Ochsenbein, S.T., Smith, A.L., Echodu, D.C., Robinson, B.H., Gamelin, D.R.: Hyperfine coupling in colloidal n-type ZnO quantum dots: effects on electron spin relaxation. J. Phys. Chem. C **114**(34), 14467–14472 (2010)

154. Lee, S.-Y., Paik, S.-Y., McCamey, D.R., Yu, J., Burn, P.L., Lupton, J.M., Boehme, C.: Tuning hyperfine fields in conjugated polymers for coherent organic spintronics. J. Am. Chem. Soc. **133**(7), 2019–2021 (2011)

155. Maurer, P.C., Kucsko, G., Latta, C., Jiang, L., Yao, N.Y., Bennett, S.D., Pastawski, F., Hunger, D., Chisholm, N., Markham, M., Twitchen, D.J., Cirac, J.I., Lukin, M.D.: Room-temperature quantum bit memory exceeding one second. Science **336**(6086), 1283–1286 (2012)

156. Jeschke, G.: Distance measurements in the nanometer range by pulse EPR. ChemPhysChem **3**(11), 927–932 (2002)

157. Borbat, P.P., McHaourab, H.S., Freed, J.H.: Protein structure determination using long-distance constraints from double-quantum coherence ESR: study of T4 lysozyme. J. Am. Chem. Soc. **124**(19), 5304–5314 (2002)

158. Duckheim, M., Loss, D.: Electric-dipole-induced spin resonance in disordered semiconductors. Nat. Phys. **2**(3), 195–199 (2006)

159. Harris, C., Tinti, D., El-Sayed, M., Maki, A.: Optical detection of phosphrescent triplet state endor in zero field. Chem. Phys. Lett. **4**(6), 409–412 (1969)

160. Guerrier, D.J., Harley, R.T.: Calibration of strain vs nuclear quadrupole splitting in III–V quantum wells. Appl. Phys. Lett. **70**(13), 1739–1741 (1997)

161. Weil, J.A., Bolton, J.R.: Electron Paramagnetic Resonance: Elementary Theory and Practical Applications, 232, 2nd edn. Wiley, Hoboken, NJ, USA (2007)

162. Rieger, P.H.: Electron Spin Resonance: Analysis and Interpretation, pp. 122–126. The Royal Society of Chemistry, Cambridge, UK (2007)

163. Weil, J.A., Bolton, J.R.: Electron Paramagnetic Resonance: Elementary Theory and Practical Applications, chapter 6, 2nd edn. Wiley, Hoboken, NJ, USA (2007)

164. Pake, G.E.: Nuclear resonance absorption in hydrated crystals: fine structure of the proton line. J. Chem. Phys. **16**(4), 327–336 (1948)

165. Wasserman, E., Snyder, L.C., Yager, W.A.: ESR of the triplet states of randomly oriented molecules. J. Chem. Phys. **41**(6), 1763–1772 (1964)

166. Jeschke, G.: Determination of the nanostructure of polymer materials by electron paramagnetic resonance spectroscopy. Macromol. Rapid Commun. **23**(4), 227–246 (2002)

167. Abragam, A., Bleaney, B.: Electron Paramagnetic Resonance of Transition Ions, pp. 152–155. Oxford University Press, London, UK (1970)

168. Gromov, I., Schweiger, A.: Multiphoton resonances in pulse EPR. J. Magn. Reson. **146**(1), 110–121 (2000)

169. Rabi, I.I.: Space quantization in a gyrating magnetic field. Phys. Rev. **51**(8), 652–654 (1937)

170. Torrey, H.: Transient nutations in nuclear magnetic resonance. Phys. Rev. **76**(8), 1059–1068 (1949)

171. Yoshie, T., Scherer, A., Hendrickson, J., Khitrova, G., Gibbs, H.M., Rupper, G., Ell, C., Shchekin, O.B., Deppe, D.G.: Vacuum Rabi splitting with a single quantum dot in a photonic crystal nanocavity. Nature **432**(7014), 200–203 (2004)

172. Rabi, I.I., Ramsey, N., Schwinger, J.: Use of rotating coordinates in magnetic resonance problems. Rev. Mod. Phys. **26**(2), 167–171 (1954)

173. Hagelstein, P.L., Senturia, S.D., Orlando, T.P.: Introductory Applied Quantum and Statistical Mechanics, pp. 307–315. Wiley, Hoboken, NJ, USA (2004)

174. Gliesche, A., Michel, C., Rajevac, V., Lips, K., Baranovskii, S.D., Gebhard, F., Boehme, C.: Effect of exchange coupling on coherently controlled spin-dependent transition rates. Phys. Rev. B **77**(24), 245206 (2008)

175. Ayabe, K., Sato, K., Nishida, S., Ise, T., Nakazawa, S., Sugisaki, K., Morita, Y., Toyota, K., Shiomi, D., Kitagawa, M., Takui, T.: Pulsed electron spin nutation spectroscopy of weakly exchange-coupled biradicals: a general theoretical approach and determination of the spin dipolar interaction. Phys. Chem. Chem. Phys. **14**(25), 9137–9148 (2012)

176. Limes, M.E., Wang, J., Baker, W.J., Lee, S.-Y., Saam, B., Boehme, C.: Numerical study of spin-dependent transition rates within pairs of dipolar and exchange coupled spins with s = 1/2 during magnetic resonant excitation. Phys. Rev. B 87, 165204 (2013)

177. Glenn, R., Limes, M.E., Saam, B., Boehme, C., Raikh, M.E.: Analytical study of spin-dependent transition rates within pairs of dipolar and strongly exchange coupled spins with s = 1/2 during magnetic resonant excitation. Phys. Rev. B 87, 165205 (2013)
178. Astashkin, A.V., Schweiger, A.: Electron-spin transient nutation: A new approach to simplify the interpretation of ESR spectra. Chem. Phys. Lett. **174**(6), 595–602 (1990)
179. Rajevac, V., Boehme, C., Michel, C., Gliesche, A., Lips, K., Baranovskii, S.D., Thomas, P.: Transport and recombination through weakly coupled localized spin pairs in semiconductors during coherent spin excitation. Phys. Rev. B **74**(24), 245206 (2006)
180. Glenn, R., Baker, W.J., Boehme, C., Raikh, M.E.: Analytical description of spin-Rabi oscillation controlled electronic transitions rates between weakly coupled pairs of paramagnetic states with s = 1/2. Phys. Rev. B 87, 155208 (2013)
181. Weil, J.A., Bolton, J.R.: Electron Paramagnetic Resonance: Elementary Theory and Practical Applications, 2nd edn., pp. 312–315. Wiley, Hoboken, NJ, USA (2007)
182. Hahn, E.L.: Spin echoes. Phys. Rev. **297**(1946), 580–594 (1950)
183. Mims, W., Nassau, K., McGee, J.: Spectral diffusion in electron resonance lines. Phys. Rev. **123**(6), 2059–2069 (1961)
184. Rowan, L., Hahn, E., Mims, W.: Electron-spin-echo envelope modulation. Phys. Rev. **137**(1A), A61–A71 (1965)
185. Mims, W.B.: Envelope modulation in spin-echo experiments. Phys. Rev. B **5**(7), 2409–2419 (1972)
186. Mims, W.B.: Amplitudes of superhyperfine frequencies displayed in the electron-spin-echo envelope. Phys. Rev. B **6**(9), 3543–3545 (1972)
187. Breiland, W., Harris, C., Pines, A.: Optically detected electron spin echoes and free precession in molecular excited states. Phys. Rev. Lett. **30**(5), 158–161 (1973)
188. van Oort, E., Glasbeek, M.: Optically detected low field electron spin echo envelope modulations of fluorescent NV centers in diamond. Chem. Phys. **143**, 131–140 (1990)
189. Hoehne, F., Dreher, L., Huebl, H., Stutzmann, M., Brandt, M.: Electrical detection of coherent nuclear spin oscillations in phosphorus-doped silicon using pulsed ENDOR. Phys. Rev. Lett. **106**(18), 187601 (2011)
190. Dikanov, S.A., Tsvetkov, Y.D.: Electron Spin Echo Envelope Modulation (ESEEM) Spectroscopy. CRC, Boca Raton, FL, USA (1992)
191. Hilton, D., Tang, C.: Optical orientation and femtosecond relaxation of spin-polarized holes in GaAs. Phys. Rev. Lett. **89**(14), 146601 (2002)
192. Yu, P., Beard, M.C., Ellingson, R.J., Ferrere, S., Curtis, C., Drexler, J., Luiszer, F., Nozik, A.J.: Absorption cross-section and related optical properties of colloidal InAs quantum dots. J. Phys. Chem. B **109**(15), 7084–7087 (2005)
193. Abragam, A., Bleaney, B.: Electron Paramagnetic Resonance of Transition Ions, chapter 10. Oxford University Press, London, UK (1970)
194. Merkulov, I.A., Efros, A.L., Rosen, M.: Electron spin relaxation by nuclei in semiconductor quantum dots. Phys. Rev. B **65**(20), 205309 (2002)
195. Syperek, M., Yakovlev, D., Yugova, I., Misiewicz, J., Sedova, I., Sorokin, S., Toropov, A., Ivanov, S., Bayer, M.: Long-lived electron spin coherence in CdSe/Zn(S,Se) self-assembled quantum dots. Phys. Rev. B **84**(8), 085304 (2011)
196. Feng, D.H., Li, X., Jia, T.Q., Pan, X.Q., Sun, Z.R., Xu, Z.Z.: Long-lived, room-temperature electron spin coherence in colloidal CdS quantum dots. Appl. Phys. Lett. **100**(12), 122406 (2012)
197. Gupta, J., Awschalom, D., Peng, X., Berezovsky, J., Lau, W., Ghosh, S., Stephens, J., Stern, N.: Spin coherence in semiconductors. In: Flatté, M., Tifrea, I. (eds.) Manipulating Quantum Coherence in Solid State Systems. Volume 244 of NATO Science Series, pp. 130–170. Springer, Dordrecht, Netherlands (2007)
198. Ochsenbein, S.T., Gamelin, D.R.: Quantum oscillations in magnetically doped colloidal nanocrystals. Nat. Nanotechnol. **6**(2), 112–115 (2011)

Chapter 2
Experimental Methods

2.1 Experimental Considerations of pODMR

For the development of the pODMR studies outlined in Chaps. 3 and 4, it is important to fully understand the technical nature of the experiments performed. To this end, a detailed overview of several of the salient experimental issues is discussed in this section. First a walk-through of the technical setup is given, which outlines some of the more critical features of physically implementing these experiments. Also critical to any measurement is the knowledge of how a dynamic signal becomes digitized, which then determines how those data are later analyzed. A description of how the resonantly transient response in photoluminescence is captured and digitized is given, enabling a meaningful scaling of data that are acquired with the Bruker Elexsys E580.[1] Calibration of this spectrometer's input analog-to-digital converter (ADC) is also crucial for scaling data correctly, and is also discussed. Finally, since the ODMR observable is a resonant change in photoluminescence intensity, it is imperative that the researcher have complete knowledge of the emitting species which are supported by the material under investigation. To this end, the need for spectral selection of the desired emission band is demonstrated with an example given of the confusion which can result from incomplete knowledge of a material's minority emission channels.

2.1.1 Technical Implementation

The prospect of conducting an optical experiment under spin-resonant conditions inherently means that a marriage of two traditional types of optical probes must be performed: at one end, a traditional photoluminescence setup and at the other,

---

[1] *Bruker BioSpin Corp.*; Billerica, MA, USA; X-band EPR spectrometer.

K. van Schooten, *Optically Active Charge Traps and Chemical Defects in Semiconducting Nanocrystals Probed by Pulsed Optically Detected Magnetic Resonance*, Springer Theses, DOI 10.1007/978-3-319-00590-4_2, © Springer International Publishing Switzerland 2013

a somewhat conventional electron spin resonance setup. In order to operate these systems in concert, some care must be taken to make one compatible with the other. Discussed here is how optical excitation was given access to the highly space-constrained environment of the spin-resonance chamber and how photoluminescence was collected, amplified, and coordinated with the timing of nanosecond-scale microwave pulses. This is schematically outlined in Fig. 2.1.

The sample space of the ESR system primarily consists of a low-Q, dielectric FlexLine resonator,[2] which supports, and then quickly dampens, the applied microwave field pulse. The cylindrical dimensions of the resonator are defined by a height of approximately 7 cm and a radius of 3 cm. The resonant volume is much smaller and is defined instead by the distribution of homogeneous magnetic fields that are induced by the applied microwave field and concentrated at the center of the cylindrical volume. The field distribution can be considered negligible over a cylindrical volume roughly defined by an 8 mm height and a 3 mm radius. This smaller resonant volume sets the limit on sample geometry. Since the entire resonator is housed within a Helium-4 flow cryostat,[3] stringent constraints on any options for optical access are imposed. The most practical access point is through a 7 mm port which is meant to carry a standard Bruker sample rod.[4] In the work presented here, the sample rod was customized by the author in order to act as a port for a fiber bundle. This fiber bundle is comprised of one excitation fiber and six collection fibers. In designing this fiber bundle, options for larger numerical aperture fibers[5] were chosen in order to maximize photoluminescence collection efficiencies. Since the fibers are composed of a paramagnetically inert glass, inserting them directly into the FlexLine resonator does not alter the modal distribution of magnetic fields.

Samples are prepared by drop-casting a nanoparticle solution into a Teflon "bucket" (details of this process are outlined further in Sect. 2.4). This optically and paramagnetically inert bucket is placed in the bottom of a standard 4 mm diameter quartz EPR tube.[6] A collar[7] attached to the FlexLine resonator securely holds the quartz tube in place, while the Teflon bucket is designed with a diameter just large enough to allow the insertion of the fiber bundle tip. Sealing the sample rod such that it remains vacuum-hard under a differential pressure allows it to be operated within the ^{4}He cryostat under cryogenic conditions.

---

[2]*Bruker BioSpin Corp.*; Billerica, MA, USA; ER 4118X-MD5 X-band resonator.

[3]*Bruker BioSpin Corp.*; Billerica, MA, USA; ER 4118CF ^{4}He cryostat.

[4]*Bruker BioSpin Corp.*; Billerica, MA, USA; E4118130 FlexLine Sample Rod.

[5]*Thorlabs, Inc.*; Newton, NJ, USA; excitation fiber:BFH22-550; collection fibers: BFL22-365.

[6]*Wilmad-LabGlass*; Vineland, NJ, USA; part number: 707-SQ-250M.

[7]*Bruker BioSpin Corp.*; Billerica, MA, USA; E4118140 Sample Holder Set.

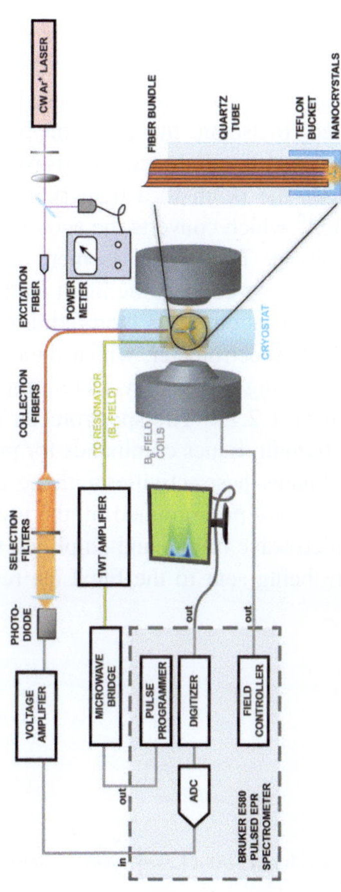

Fig. 2.1 A schematic of the pODMR experimental setup. Only major elements which are crucial to understanding the implementation of this experiment are included

Optical excitation is performed with a continuous wave (cw) Argon-ion laser.[8] A narrow-band spectral clean-up filter[9] is used to isolate the desired laser line (normally 458 nm for these studies). Laser power is measured with a photodiode[10] and is controlled with a continuously-variable, neutral density filter.[11] The free-space laser is fed into the excitation fiber with a fiber coupler,[12] leading to optical excitation of the nanocrystals. Resultant optical emission is then collected by the remaining fibers in the bundle, which are output to the focal point of a collimating lens. The purpose of this collimating lens is to preserve the collected light intensity while the beam passes through a series of optical filters. Several filters may be employed in order to isolate and monitor the spin-resonant response of a particular emission band. The first filter in this series is always an ultrasteep long-pass edge filter,[13] designed to eliminate any residually collected laser emission. Subsequent filters are then used, as needed, to isolate the emission band of interest. Once appropriately filtered, the remaining photoluminescence is passed through a second collimating optic, which focuses the beam to a tight point. Placed at this focal point is a low-noise photodiode,[14] which converts the acquired photoluminescence intensity into an amplified voltage. This signal is then further amplified by a voltage amplifier,[15] which capacitively-couples the input so that only the transient response is amplified and output. This output is then fed directly into the internal digitizer of the Bruker E580 EPR spectrometer, which then parses the data to an external control computer for recording and display. Details on the digitizer unit and its ADC front-end are given in Sect. 2.1.3. The spectrometer system also controls the static magnetic field ($\mathbf{B}_0$) strength, issues commands for predefined microwave pulse ($\mathbf{B}_1$) sequences, and coordinates these activities with the signal acquired by the digitizer. Desired pulse sequences are programmed by the user and are ultimately implemented by an external microwave bridge and amplified by a traveling-wave-tube (TWT) amplifier[16] before being sent to the FlexLine resonator for resonant excitation of the nanocrystals.

---

[8] *Coherent, Inc.*; Santa Clara, CA, USA; Innova 90-4 Laser System, supplied by *Laser Innovations*, Santa Paula, CA, USA.

[9] *Semrock, Inc.*; Rochester, NY, USA; 459.9 nm MaxLine, part number:LL01-458-12.5.

[10] *Coherent, Inc.*; Santa Clara, CA, USA; FieldMaxII with diode sensitivity range of 400–1,060 nm.

[11] *Thorlabs, Inc.*; Newton, NJ, USA; part number: NDC-50C-2M.

[12] *Thorlabs, Inc.*; Newton, NJ, USA; part number: F220SMA-A.

[13] *Semrock, Inc.*; Rochester, NY, USA; 458 nm RazorEdge filter, part number:LP02-458RU-25.

[14] *FEMTO Messtechnik GmbH*; Berlin, Germany; Si Photoreceiver, part number: LCA-S-400K-SI.

[15] *Stanford Research Systems, Inc.*; Sunnyvale, CA, USA; SR560 Low Noise Preamplifier.

[16] *Applied Systems Engineering, Inc.*; Fort Worth, TX, USA; Model 117 1kW X-band amplifier.

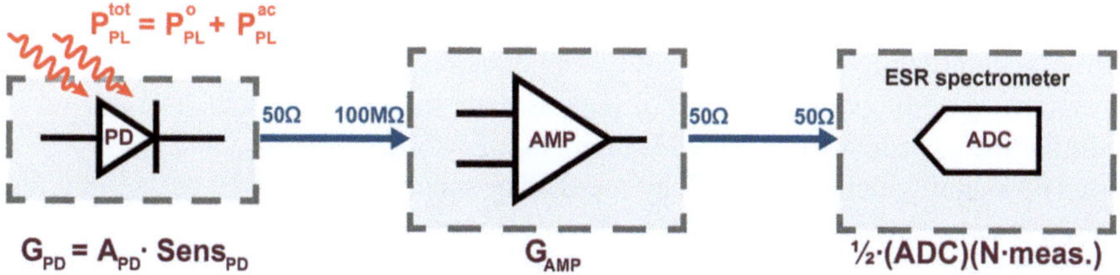

Fig. 2.2 Instrumental scaling factor for ESR spectrometer data. The general manner in which the measured change in PL intensity is converted to a voltage and then digitized is schematically shown. Signal acquisition of a transient event in photoluminescence is primarily accomplished with a low-noise photodiode, voltage amplifier, and the spectrometer ADC input stage, each of which treat their input signal in a unique manner. Input and output impedances for each are also given

2.1.2 Scaling of ESR Spectrometer Data

Under continuous-wave excitation of the nanocrystals, a steady-state photoluminescence power (P_{PL}^{o}) is maintained. Under magnetic resonance of an optically active charge carrier spin, the large static photoluminescence intensity will be slightly modulated by the transient response of the system (P_{PL}^{ac}) to give a total emitted power $P_{PL}^{tot} = P_{PL}^{o} + P_{PL}^{ac}$. This photoluminescence signal is focused to a Si-photodiode where it is converted to an analog voltage, as shown in Fig. 2.2. The total gain response of the photodiode is defined by the product of transimpedance gain (A_{PD}) and spectral sensitivity ($SENS_{PD}$) of the Si device, which means that power conversion is dependent on the emission spectrum being collected. Since the PL has both a steady-state and transient component, so will the voltage signal from the photodiode, $V_{PD}^{tot} = V_{PD}^{o} + V_{PD}^{ac}$. Or, in terms of the input and gain, $V_{PD}^{tot} = A_{PD} \cdot SENS_{PD} \left(P_{PL}^{o} + P_{PL}^{ac} \right)$.

Since the input stage of the amplifier has a very high impedance ($100\,M\Omega$), there is virtually no loss in signal amplitude before gain (G_{AMP}) is applied. However, there is $50\,\Omega$ impedance matching between the amplifier output and the ADC input of the spectrometer's digitizing unit. This effectively acts as a voltage divider, cutting the signal's voltage amplitude in half (hence the factor of $\frac{1}{2}$ in Fig. 2.2). The ADC itself is specified to be 8-bit (256 steps/V) and is a direct front-end to the digitizing unit with no additional signal processing in between. Calibration of the ADC is a very important and necessary step to correctly scale raw data and is covered in Sect. 2.1.3. Recent calibration shows that the ADC actually digitizes data on a slope of 220 steps/V.

In order to increase the signal-to-noise ratio of a measurement, extensive averaging is normally employed. Experimentally, two types of averaging are generally used. One successively sums acquired signals over the course of a large number of microwave shots; typically $N_{shots} \approx 16,000$ with a repetition time of 100–$1,000\,\mu$s between each shot. This is normally done while keeping all other parameters fixed (i.e. external B_0 field, laser power, temperature, etc.). This fast repetition is intended to average away any short-term noise. Noise which exists over longer time scales is reduced by repeating the above N_{shots} after some longer wait time, accumulating many sets of equivalent data, $N_{accum.}$; this typically occurs after performing some parameter sweep, such as external magnetic field. The full set of similar measurements performed is then $N_{meas.} = N_{shots} \cdot N_{accum.}$.

The signal which is digitized by the Bruker spectrometer can be scaled in terms of the measurable parameter, P_{PL}^{ac},

$$Raw\ Data = \left(\frac{1}{2} \cdot 220 \frac{steps}{V} \cdot N_{shots} \cdot N_{accum.} \right) \cdot (G_{AMP}) \cdot (A_{PD} \cdot SENS_{PD}) \cdot \left(P_{PL}^{ac} \right).$$

Although it is technically possible to scale the data absolutely in terms of changes in PL power (i.e. nanowatts), this is experimentally difficult due to the spectral sensitivity, $SENS_{PD}$, of the Si photodiode. It would then be necessary to project the acquired emission spectrum against the spectral sensitivity curve of the diode in order to determine the actual value of $SENS_{PD}$. Thus, while certainly possible, this process can be cumbersome and time consuming. A convenient alternative is to quote the observable in terms of a percent change in photoluminescence intensity (or power),

$$\%Diff = \frac{(P_{PL}^{o} + P_{PL}^{ac}) - P_{PL}^{o}}{P_{PL}^{o}} \cdot 100$$

$$= 2 \cdot 100 \cdot \left(\frac{1V}{220 steps} \right) \cdot \left(\frac{1}{V_{PD}^{o}} \right) \cdot \left(\frac{1}{G_{AMP}} \right) \cdot \left(\frac{RawData}{N_{shots} \cdot N_{accum.}} \right).$$

Because $SENS_{PD}$ cancels from the relation, it is much more straightforward to simply measure V_{PD}^{o} from the photodiode while under nonresonant conditions than to measure the emission spectrum and assess the approximate value of $SENS_{PD}$.

2.1.3 Signal Input Calibration for the ESR Spectrometer

As mentioned above in Sect. 2.1.2, a knowledge of how the spectrometer's internal ADC divides the input voltage into discrete digital steps is crucial to the correct scaling of the raw data it acquires. This is especially so for the studies described in Chaps. 3 and 4, where nonstandard spectrometer inputs were utilized. Since the Bruker E580 spectrometer system being used here is a self-contained, purpose-built unit designed to perform standard ESR measurements, the internal operations of this machine are somewhat opaque to the user. Here, a general calibration procedure will be outlined, ensuring that at least some of the internal workings of this spectrometer will be transparent for future users.

According to the Bruker E580 User's Manual, ESR signals are input into a quadrature detection stage, which mainly consists of an input preamp, power splitter, phase shifter, and signal mixers. Outputs from the two mixers are fed into additional amplifiers and sent out for digitization. An extremely important point is that the entire quadrature detection stage is bypassed during ODMR measurements.

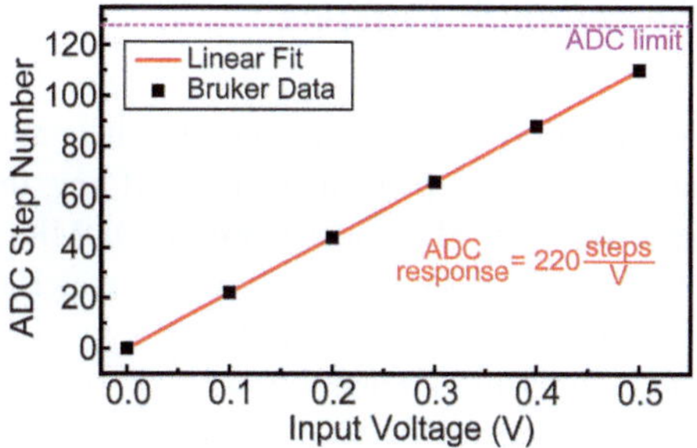

Fig. 2.3 Calibration results for signal input ADC of ESR spectrometer. The spectrometer input ADC is specified by the manufacturer as being an 8-bit digitizer giving 256 steps/V over a range of ± 0.5 V. Since the number of divisions per volt input is an important factor for correctly scaling data obtained with this machine, it is necessary to verify the operation of this unit. Shown are the results of such a verifying measurement, demonstrating that deviation from specified operational values can be significant

The outputs of the quadrature stage are routed to a digitizing unit with an ADC front-end. This is actually where the ODMR signal directly enters the spectrometer, which invalidates certain user input parameters such as Video Gain and Bandwidth since the signal experiences no additional filtering or amplification.

Since the ODMR signal is patched directly into the ADC, it then becomes important to validate the calibration of this element. In consulting the limited schematics contained within the Bruker E580 User's Manual, this ADC is quoted as having an 8-bit resolution with 256 steps/V over a range of ± 0.5 V. The veracity of this statement can, and should, be checked by the user since the operation of this element directly affects how an external signal is digitally scaled. To do this requires applying a dc voltage to the ADC input, sweeping the voltage over several values within the ADC operating range while simultaneously monitoring the reading given by the spectrometer in the computer display. Within the spectrometer control software, a "dummy" experiment needs to be set up, consisting of a single acquisition (e.g. $N_{meas.} = N_{shots} \cdot N_{accum.} = 1$) in order to take this reading. After a voltage sweep has been made, a simple linear fit to the ADC output is plotted as a function of applied voltage, revealing the actual sampling and quantization process. An example of such a calibration can be seen in Fig. 2.3, where the digitizer unit is actually found to operate with a dependence of 220 steps/V. The discrepancy of this with manufacturer specifications results in an error in measured $\%Diff$ of -14.0625%. Most ADCs are internally calibrated through the use of a reference voltage, and any deviation in the value of this reference voltage would cause a drift in the ADC's calibration. This may be the cause for the measured deviation from manufacturer specifications.

2.1.4 Spectral Selection in Practice

In any systematic investigation of a physical system, it is absolutely necessary to first understand what competing dependencies exist, if any, that may contribute to the observable of that experiment [1]. For the case of ODMR, this primarily means performing some form of optical spectroscopy on the material of interest in order to develop an understanding of the emission channels involved. This is important for ODMR since optical emission intensity is the experimental observable and, if there are multiple emitting species or sites within a material, then there may be multiple classes of optically active carriers contributing to the observable. Common dual-emission channels are usually excitonic band edge emission coexisting with donor-acceptor emission [2], or prompt fluorescence followed by weakly allowed phosphorescence emission [3]. Discussed at length in Chaps. 3 and 4 are ODMR studies conducted on materials with competing emission channels (primarily band edge versus donor-acceptor emission), where each channel emits at a separate energy level. The energetically separated emission allows the investigator to spectrally select the photoluminescence band of interest as a probe of only those paramagnetic states that play a role in that channel's emission process. Such selection can be easily established by placing optical isolation filters in the pathway of photoluminescence collection optics, allowing only the emission channel of interest to illuminate the photodiode (see Fig. 2.1 for placement of selection filters).

It is often the case, though, that materials with multiple emitting states have a single channel with very large oscillator strength and therefore dominates any simple cw spectrum that is acquired. A spectrometer outfitted with a camera possessing a very large dynamic range and the ability to take time-gated spectra is of great use in such a case; such a system proved invaluable to the studies outlined in Chaps. 3 and 4. The operations of this system are discussed further in Sect. 2.2.

As an example of the importance of spectral selection in ODMR, Fig. 2.4 demonstrates the confusion which can arise from neglecting to account for the multiple emitting states found in some materials. The material that is shown here is the CdSe/CdS nanotetrapod, which is the subject of Chap. 3 and is described further in Sect. 2.3. The CdS arms of these nanocrystals have a very large optical absorption cross-section as well as a larger band gap (by ~0.7 eV), as compared to the much smaller CdSe core. This means that nearly all optical excitation takes place in the CdS arms, but the excitation energy is efficiently transferred to the core where emission takes place. Panel (a) of Fig. 2.4 shows the emission spectrum of the tetrapods, where the only emission channel visible on a linear intensity scale is that at 2.0 eV (i.e. the CdSe core). For the ODMR measurements shown in panel (b), a 458 nm ultrasteep long-pass edge filter was used to collect emission from the region marked with green in (a). Simply checking the dependence of resonance lineshape on laser excitation power reveals a curious "morphing" progression of the ODMR lineshape; at low laser power, the spectrum is primarily dominated by three PL enhancement resonances, while at higher excitation densities, the lineshape becomes quite complicated, with obvious photoluminescence quenching

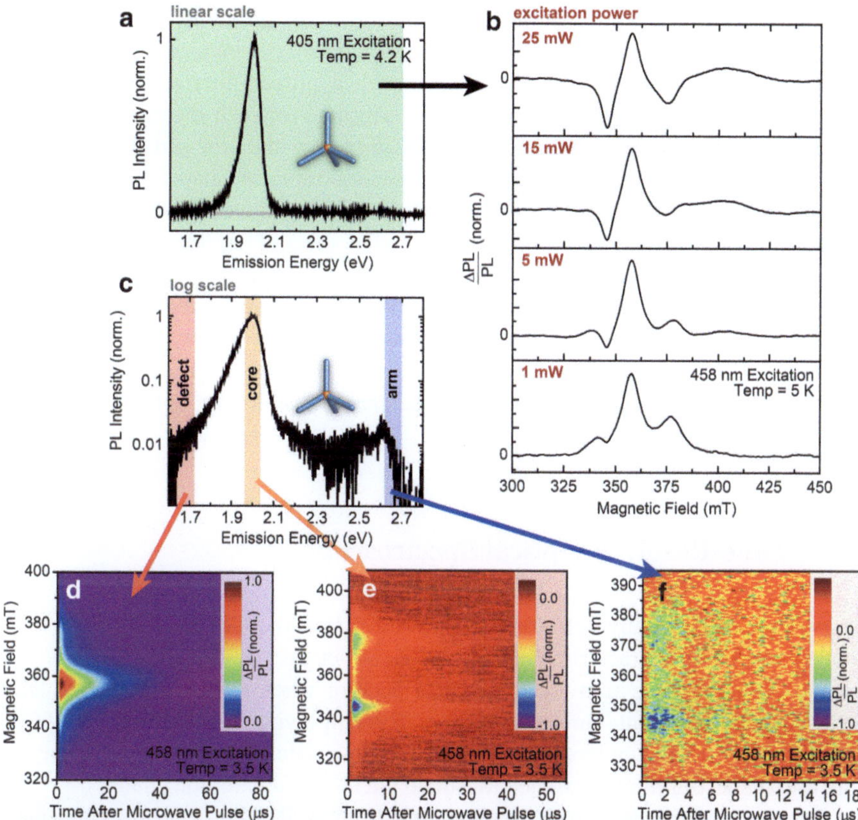

Fig. 2.4 ODMR signal convolution from multiple emission channels and the need for spectral selection. Panel (**a**) shows the emission spectrum of CdSe/CdS nanotetrapods. The CdS arms of these nanocrystals have a very large optical absorption cross-section as compared to the CdSe core, meaning that nearly all optical excitation takes place there. Excitation energy is transferred to the CdSe core, where emission takes place at roughly 2.0 eV. In assuming that the core emission is the only emission channel of these nanocrystals, performing an ODMR experiment presents some challenges due to a partial ignorance of the system under study. Panel (**b**) demonstrates this in the "morphing" of resonance structure with respect to a change in applied laser power. The same optical spectrum of (**a**) is shown again on a log-intensity scale in (**c**), revealing nearly insignificant emission from the CdS arms [4] and deep-level defects [5]. Choosing optical isolation filters to select each of these bands (*colored regions*) and repeating the ODMR measurement allows one to probe only the spin-resonant carriers involved with each emitting species (**d**)–(**f**)

features. Such a change to the lineshape is not expected to occur for an ensemble system which simply experiences an increase in the population of paramagnetic centers. This somewhat odd behavior firmly indicates that there are multiple spin-resonant species contributing to the overall emission from these nanocrystals, which then leads to a convolution of resonance lineshapes. The presence of multiple emission centers in the tetrapods is confirmed in panel (c), where the same spectrum given in (a) is instead shown on a logarithmic PL intensity scale.

Three emission bands are identified: residual CdS arm emission [4]; the dominant CdSe core emission; and the onset of deep-level defect emission. Each of these photoluminescence bands can be isolated with a proper filter set (indicated by color bands in (c)), allowing ODMR to be performed for each segregated emission channel (panels (d)–(f)). The resonances observed by the arm and core emission are qualitatively very similar, which is the general topic of Chap. 3. Both of these resonances are distinct, however, from the resonance obtained from the deep-level defect emission, which is the topic of Chap. 4. The spectral lineshape dependence on excitation power seen in (b) represents a convolution of these three spin-resonance features (d)–(f), with each emission channel increasing its paramagnetic population density at different rates in response to changes in optical density. Experimentally confirmed, but not shown here, is that the resonance spectrum of each of these emission bands (d)–(f) remains unchanged, regardless of the applied excitation power.

2.2   Time-Resolved Optical Spectroscopy

Imperative to spectrally selected ODMR is a direct knowledge of the material's optical emission characteristics. As was shown in Sect. 2.1.4, unambiguous resonance structure is extremely difficult to acquire when the observable contains multiple spin-resonant emitting species. By employing some form of optical spectroscopy, a clear understanding of the emitting species can be gained, enabling the use of spectral filters designed to isolate the intended emission channel.

For the work discussed in Chaps. 3 and 4, an Andor SR-303i[17] spectrometer with an attached time-gated, intensified charge-coupled device (Andor iStar DH720 ICCD[18]) was used. This experimental setup is fully described in the Ph.D. thesis of Su Liu [6]. The use of this instrument made available a great deal of knowledge about the energetic nature of the nanocrystals being studied. The dynamic range of the ICCD camera is well over eight orders in magnitude, allowing for the detection of extremely weakly emitting states. Since the ODMR observable is a differential change in photoluminescence intensity, usually from 0.001 to 1.0% of total PL yield, even the most subtle emission process can have a dramatic effect on the measurement. By obtaining high-quality optical spectra, complications arising from such weak emitters can be avoided.

The requirements for an optically active carrier to become observable in a pODMR experiment are two-fold: first is that the carrier experience a spin-lifetime at least on the order of $T_1 \approx 10\,ns$; secondly, the carrier lifetime must also be on the order of, or exceed, $10\,ns$. The first condition must be checked under

---

[17]*Andor Technology PLC*; Belfast, Ireland; Imaging spectrometer with UV-Vis mirror and 150 l/mm (1.57 nm resolution) ruled grating.

[18]*Andor Technology PLC*; Belfast, Ireland; with 18 mm generation 3 intensifier tube.

spin-resonance conditions, while the second can be investigated with the time-gated ICCD spectrometer just described. The gating operation for this ICCD extends from 2 ns to 2 s, with subnanosecond step resolution. The timing of this camera is managed by an internal digital delay generator which is triggered by an external pulsed laser (355 nm)[19] operating at variable frequency (single shot to 2 kHz). This laser is also used as a photoluminescence excitation source for the material under investigation. In this way, optical lifetimes and decay spectra can be obtained. An example of this is discussed further in Sect. 3.6.2.

2.3 Nanocrystal Materials

The studies discussed in Chaps. 3 and 4 revolve around the optically active electronic states of two types of nanocrystals, namely, the CdS nanorods and the CdSe/CdS tetrapods. In order to provide some perspective on these studies, some of the material attributes of these nanocrystals will be discussed here. A special note is made that these materials were fully synthesized and graciously made available by Dmitri V. Talapin of the University of Chicago and his graduate student, Jing Huang.

As their common moniker implies, colloidal nanocrystals are synthesized [7] via precipitation from chemical precursor materials which are held in colloidal suspension within an acceptable solvent. Generally, a stoichiometric amount of precursor materials (typically Cd- and S-bearing complexes, rather than bare elements) is combined with organic surfactants in a temperature-controlled solvent bath. As the temperature is raised, the precursor compounds dissociate, allowing nucleation centers to form. By controlling the rate kinetics of crystalline growth (i.e. through bath temperature or surfactant concentration) as well as the growth time, precise control over nanocrystal size, crystalline phase, *and* geometry has been demonstrated [8, 9]. For instance, the CdS nanorod ensemble has a highly homogeneous size distribution, each having dimensions of ~6 × 30 nm, and a well-defined crystalline phase (wurtzite).

These two types of nanocrystal (rods and tetrapods) have related material properties since the tetrapod form is essentially composed of four CdS nanorods (~6 × 30 nm) attached to a small (4 nm) CdSe core in a tetrahedral geometry. Four facets of zincblend CdSe serve as nucleation points of wurtzite-CdS arm growth, where the lattice mismatch between materials is ~3.8% [9]. The interface of these two materials yields a quasi-type II heterostructure, where the conduction bands nearly align and the CdSe valence band is lower by roughly 0.7 eV [10, 11]. Important here is that the much larger extent of CdS material, as compared to the CdSe core, gives the tetrapod arms a correspondingly larger optical absorption cross-section for photon energies above the CdS band gap. Nearly all optical

---

[19]*CryLas GmbH*; Berlin, Germany; Diode pumped passively Q-switched solid state laser, model FTSS 355-50.

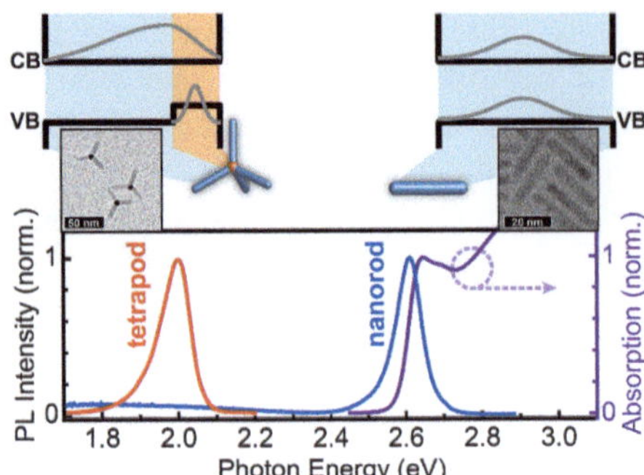

Fig. 2.5 Optical spectra characterizing the CdS nanorods and CdSe/CdS tetrapod band gaps. The tetrapod nanocrystals are composed of four CdS arms (approximately 6×30 nm) grown from a 4 nm diameter CdSe core. Optical excitation primarily takes place in the arms, due to the fact that the arms have a much larger absorption cross-section than the smaller cores [9]. Even though the CdS nanorods are practically the same dimension as a single tetrapod arm, the two sets of nanocrystals have widely separated emission energies; the nanorods emit just below the onset of CdS nanorod absorption, whereas the tetrapods absorb in their CdS arms, but emit at a nearly 0.7 eV lower energy due to the valence band offset [10, 11] of the CdSe core. An additional nanorod emission channel is also weakly visible as a broad band extending down in energy from ~2.35 eV. This represents deep-level chemical defect emission [5]. A transmission electron micrograph (TEM) of each nanoparticle species confirms the high quality of synthesis (TEM images and CdS nanorod absorption and emission spectra are kindly provided by Jing Huang)

excitation takes place in the arms, yet the core possesses the smaller band gap, so energy is efficiently transferred to the CdSe core [12–14] within ~2 ps [15]. The action of this system represents an inorganic analogue to the photosynthetic light-harvesting complexes normally encountered in organic plant life [16].

The action of this light-harvesting effect can be observed in the tetrapod structures by comparing the optical spectra of the CdS rods to that of the tetrapod heterostructure. The nanorod absorption spectrum is shown in Fig. 2.5 (violet curve), which has an onset at 2.6 eV at room temperature. Optical excitation of the nanorods results in band edge emission (blue curve) with a small Stokes shift. Nearly imperceptible on this scale is the broad deep-level chemical defect emission inherent to these nanorods [5], extending down in energy from ~2.35 eV. Alternatively, optical excitation of the tetrapod nanocrystals above the CdS band gap eventuates in emission at a lower energy (orange curve). The difference between these two emission energies equals the magnitude of valence band offset between the CdS and CdSe nanomaterials [10, 11]. The primary topic of Chap. 3 is how the control of spin-states existing in the tetrapod arms can be utilized to exert some measure of coherent control over the general light-harvesting process for this form of nanocrystal.

2.4 Sample Preparation

For the studies that are discussed in Chaps. 3 and 4, two general types of samples
were prepared: one for optical spectroscopy and one for spin-resonance spectroscopy. All samples ultimately derived from nanocrystal mother solutions, which
were suspensions in toluene (concentrations unknown) provided by Jing Huang.
These were initially drawn from to create diluted daughter solutions in order to
preserve the purity of the mother solution. All daughter solutions were diluted
by adding 10 mg/mL polystyrene[20] (also dissolved in toluene) and permanently
stored in an inert atmosphere (N_2) glovebox system.[21] The purpose of blending
the nanocrystals with polystyrene is to help ensure that particle–particle segregation
is achieved once the solvent is removed, as well as to lock the particle ensemble in
a rigid matrix. This matrix material is ideally suited for these tasks since it is both
optically and paramagnetically inert and is also solution processable.

For optical studies, this solution is normally drop-cast onto a standard glass
microscope coverslip,[22] which is used as a substrate, and then placed on a 50 °C
hot plate for 10 min to evaporate any remaining solvent. The resultant film is
approximately 1 μm thick. These substrates are cleaned to a high degree before
usage by submitting them to a sequence of 10 min ultrasonic baths in acetone, iso-
propanol, ultra-pure water, and a second, final bath in ultra-pure water. Ultimately,
the substrates are mounted to a vertical cold-finger cryostat[23] using a thermally
conductive silver paste. The small thickness of the coverslips (0.2 mm) also aids in
efficient heat transport from the surface containing the nanoparticles. In order to
protect against the accidental collection of auto-fluorescence from the silver paste
adhesive, a roughly 100 nm thick aluminum mirror is deposited on the back of the
coverslip by thermal evaporation before starting the cleaning procedure. At this
point, time-resolved optical spectroscopy can be performed on the nanocrystal
ensemble, as outlined in Sect. 2.2.

Spin resonance studies require a different sample geometry due to the size
restrictions imposed by the microwave resonator. Here, a mounting scheme compatible with the fiber optics used for optical excitation and collection must be
employed, while also maintaining a paramagnetically inert environment. In this
case, a small Teflon "bucket" was fabricated, which contains the nanocrystal sample
and is also able to accept the fiber bundle tip (see Fig. 2.1 for a schematic view of this
description). Teflon material is chosen since it is both optically and paramagnetically
inert, but is also stable against cryogenic cycling. Fabrication is carried out from a
small-diameter Teflon rod which is mechanically trimmed with a lathe to the proper
dimensions (approximately a 4 mm height and a 3 mm diameter). As in the optical

---

[20] *Sigma-Aldrich*; USA; polystyrene, purity >99.99%.

[21] *M. Braun, Inc.*; Stratham, NJ, USA; custom configuration.

[22] *Carolina Biological Supply*; Burlington, NC, USA; 1 cm square, 0.2 mm thick.

[23] *R.G. Hansen & Associates*; Santa Barbara, CA, USA; closed-cycle He cryostat model: DE-202.

samples, the nanocrystal solution is drop-cast into the Teflon bucket and annealed at 50 °C for 10 min to evaporate any residual solvent, resulting in a ~1 μm thick polystyrene block. The sample is then loaded into the bottom of a standard EPR quartz tube, which also accepts the fiber bundle. The standard sample rods for the Bruker spectrometer are designed to accommodate EPR quartz tubes of this type, facilitating sample mounting with this scheme. With the apparatus and sample prepared, the spin resonance experiments described in Chaps. 3 and 4 can proceed.

References

1. Davies, J.J.: Energy transfer effects in ODMR spectra: a possible source of misinterpretation. J. Phys. C **16**(23), L867–L871 (1983)
2. Davies, J.J.: ODMR studies of recombination emission in II-VI compounds. J. Cryst. Growth **72**(1–2), 317–325 (1985)
3. Chaudhuri, D., Wettach, H., van Schooten, K.J., Liu, S., Sigmund, E., Höger, S., Lupton, J.M.: Tuning the singlet-triplet gap in metal-free phosphorescent π-conjugated polymers. Angew. Chem. **49**(42), 7714–7717 (2010)
4. Lutich, A.A., Mauser, C., Da Como, E., Huang, J., Vaneski, A., Talapin, D.V., Rogach, A.L., Feldmann, J.: Multiexcitonic dual emission in CdSe/CdS tetrapods and nanorods. Nano Lett. **10**(11), 4646–4650 (2010)
5. Chestnoy, N., Harris, T.D., Hull, R., Brus, L.E.: Luminescence and photophysics of cadmium sulfide semiconductor clusters: the nature of the emitting electronic state. J. Phys. Chem. **90**(15), 3393–3399 (1986)
6. Liu, S.: Manipulation of exciton dynamics in macrycycle molecules and inorganic semiconductor nanocrystals. PhD thesis, University of Utah (2012)
7. Talapin, D.V., Lee, J.-S., Kovalenko, M.V., Shevchenko, E.V.: Prospects of colloidal nanocrystals for electronic and optoelectronic applications. Chem. Rev. **110**(1), 389–458 (2010)
8. Li, Y.-D., Liao, H.-W., Ding, Y., Qian, Y.-T., Yang, L., Zhou, G.-E.: Nonaqueous synthesis of CdS nanorod semiconductor. Chem. Mater. **10**(9), 2301–2303 (1998)
9. Talapin, D.V., Nelson, J.H., Shevchenko, E.V., Aloni, S., Sadtler, B., Alivisatos, A.P.: Seeded growth of highly luminescent CdSe/CdS nanoheterostructures with rod and tetrapod morphologies. Nano Lett. **7**(10), 2951–2959 (2007)
10. Pandey, A., Guyot-Sionnest, P.: Intraband spectroscopy and band offsets of colloidal II-VI core/shell structures. J. Chem. Phys. **127**(10), 104710 (2007)
11. Steiner, D., Dorfs, D., Banin, U., Della Sala, F., Manna, L., Millo, O.: Determination of band offsets in heterostructured colloidal nanorods using scanning tunneling spectroscopy. Nano Lett. **8**(9), 2954–2958 (2008)
12. Muller, J., Lupton, J.M., Lagoudakis, P.G., Schindler, F., Koeppe, R., Rogach, A.L., Feldmann, J., Talapin, D.V., Weller, H.: Wave function engineering in elongated semiconductor nanocrystals with heterogeneous carrier confinement. Nano Lett. **5**(10), 2044–2049 (2005)
13. Kraus, R.M., Lagoudakis, P.G., Rogach, A.L., Talapin, D.V., Weller, H., Lupton, J.M., Feldmann, J.: Room-temperature exciton storage in elongated semiconductor nanocrystals. Phys. Rev. Lett. **98**(1), 017401 (2007)
14. Mauser, C., Limmer, T., Da Como, E., Becker, K., Rogach, A.L., Feldmann, J., Talapin, D.V.: Anisotropic optical emission of single CdSe/CdS tetrapod heterostructures: evidence for a wavefunction symmetry breaking. Phys. Rev. B **77**(15), 153303 (2008)

15. Lupo, M.G., Della Sala, F., Carbone, L., Zavelani-Rossi, M., Fiore, A., Lüer, L., Polli, D., Cingolani, R., Manna, L., Lanzani, G.: Ultrafast electron-hole dynamics in core/shell CdSe/CdS dot/rod nanocrystals. Nano Lett. **8**(12), 4582–4587 (2008)
16. Herek, J.L., Wohlleben, W., Cogdell, R.J., Zeidler, D., Motzkus, M.: Quantum control of energy flow in light harvesting. Nature **417**(6888), 533–535 (2002)

Chapter 3
Spin-Dependent Exciton Quenching and Intrinsic Spin Coherence in CdSe/CdS Nanocrystals

3.1 Chapter Synopsis

Large surface-to-volume ratios of semiconductor nanocrystals cause susceptibility to charge trapping, which can modify luminescence yields and induce single-particle blinking. Optical spectroscopies cannot differentiate between bulk and surface traps in contrast to spin-resonance techniques, which in principle avail chemical information on such trap sites. Magnetic resonance detection via spin-controlled photoluminescence enables the direct observation of interactions between emissive excitons and trapped charges. This approach allows the discrimination of two functionally different trap states in CdSe/CdS nanocrystals underlying the fluorescence quenching and thus blinking mechanisms: a spin-dependent Auger process in charged particles; and a charge-separated state pair process, which leaves the particle neutral. The paramagnetic trap centers offer control of energy transfer from the wide-gap CdS to the narrow-gap CdSe, i.e. light harvesting within the heterostructure. Coherent spin motion within the trap states of the CdS arms of nanocrystal tetrapods is reflected by spatially remote luminescence from CdSe cores with surprisingly long coherence times of >300 ns at 3.5 K.

3.2 Introduction

Substantial control over the chemistry of semiconductor nanocrystals has been demonstrated in recent years while pursuing novel optoelectronic device schemes [1–3]. Shortcomings in the performance of these materials are routinely attributed to ill-defined "trap" states competing with the quantum-confined primary exciton [4]. While frequently implicated in explaining device inefficiencies [2], photoluminescence (PL) blinking [5–9], and delayed PL dynamics [4, 10] little is known about the underlying chemical nature of these deleterious states. Despite the wealth of structural and electronic information accessible in optical spectroscopy, the

K. van Schooten, *Optically Active Charge Traps and Chemical Defects in Semiconducting Nanocrystals Probed by Pulsed Optically Detected Magnetic Resonance*, Springer Theses, DOI 10.1007/978-3-319-00590-4_3, © Springer International Publishing Switzerland 2013

spin degree of freedom has received only marginal consideration as a complementary probe of semiconductor nanocrystals. Approaches pursued previously include isolation of paramagnetic centers in doped dilute magnetic semiconductor nanoparticles [11, 12]; resolving the exciton fine structure by fluorescence spectral line narrowing [13], time-resolved Faraday rotation [12, 14] or photon-echo techniques [15]; and continuous-wave optically detected magnetic resonance (ODMR), where the fluorescence is modulated under spin-resonant excitation in a magnetic field [16–19]. The latter requires stable paramagnetic centers, where the carrier's spin and energy are maintained on long timescales compared to the oscillation period of the resonantly driven spin manifold, i.e. for tens of nanoseconds under excitation in the 10 GHz (~0.3 T) range. The persistence of spin states in bulk materials comprising heavy atoms such as cadmium is largely determined by mixing due to spin-orbit coupling.

As dimensions shrink to quantum-confined regimes, spin-orbit-driven spin-mixing mechanisms can be weakened by the discretization and separation of states, giving way to the more subtle Fermi-contact hyperfine mode of spin mixing [20]. Although spin stability can be reinforced through quantum confinement, direct band edge excitons in nanocrystals typically decay within a few nanoseconds, making them unsuitable for spin-resonant manipulation. In fact, spin mixing amongst the fine-structure levels [14, 21] of excitonic states has been shown to occur within as little as a few hundred femtoseconds by means of photon-echo spectroscopy [15]. However, electronic charge-separated or "shelved" states also exist, where the excitonic constituents – either electron or hole, or both – are stored within a trap. The carriers in this case are not necessarily lost to nonradiative relaxation, but can feed back into the exciton state at a later time. A direct visualization of this phenomenon is given by the ability to store excitons in nanoparticles under an electric field [10, 22] in analogy to excitonic memory effects in coupled quantum wells [23]. These charge-separated states can repopulate the exciton, since luminescence returns in a burst following field removal [10, 22]. While qualitative information on these shelving states (which are distinct from chemical deep traps with their characteristic red-shifted emission with respect to the exciton) continues to feed the proliferation of microscopic models of quantum dot blinking [4–9,24,25], a more quantitative metrology is required to determine the nature and location of trapped charges. Such an approach is given by the highly sensitive method of *pulsed* ODMR spectroscopy, which, in principle, is capable of chemically fingerprinting even single carrier spins.

We focus on the spin dynamics in CdSe/CdS nanocrystal tetrapods since absorption and emission can be well separated spatially and energetically: at 3.1 eV (400 nm), the absorption cross-section of the CdS arms is more than 300 times greater than that of the CdSe core [1]. Emission from CdSe dominates due to the lower band gap, making the structures excellent light-harvesting systems [1, 26]. Figure 3.1 illustrates the underlying scheme. Photons are absorbed in the arm, leading to bright CdS excitons. The conduction bands of CdS and CdSe are approximately aligned, whereas a step of ~0.7 eV exists between the valence bands. We note that significant heterogeneity in the precise energetics of the heterostructure

Fig. 3.1 A schematic of spin-dependent light harvesting in CdSe/CdS tetrapods with experimental setup. (**a**) Excitons are generated in the CdS arms by light absorption. A small fraction of these excitons becomes trapped as charge-separated states, which can re-emit an exciton to the CdS band edge. The lifetime of the trapped state is sufficient to enable spin manipulation via electron spin resonance (ESR), switching the trapped carrier pair between "bright" and "dark" mutual spin configurations. Relaxation of the exciton to the CdSe core gives rise to strongly red-shifted emission. The transmission electron micrograph inset illustrates the high quality of the structures used. (**b**) Experimental setup and a representative differential PL transient as a consequence of resonant spin transition of an optically active carrier

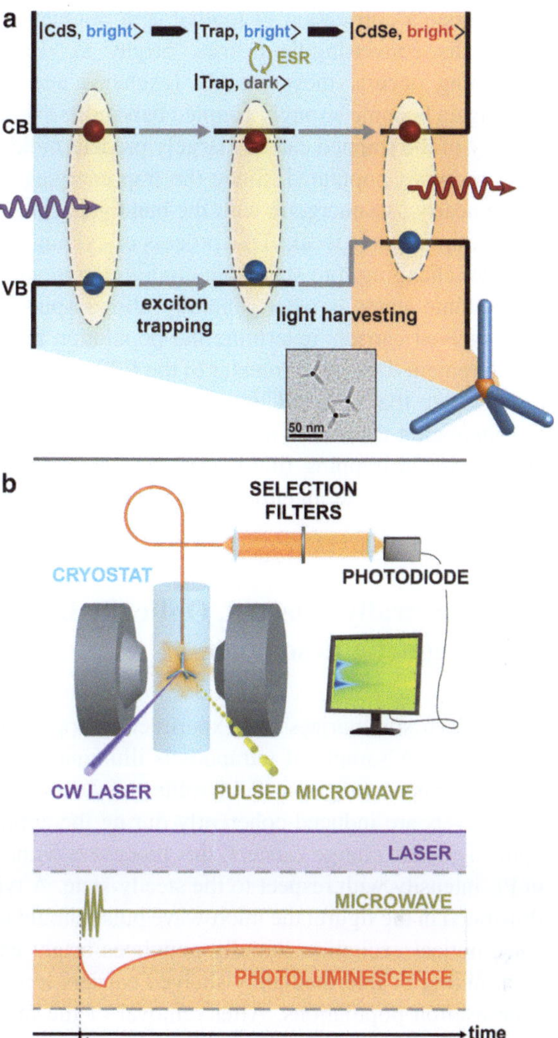

arises between single particles [26, 27]. The direct transfer of CdS excitons to CdSe is not suspected to be spin dependent since CdSe [21] and CdS [28] ground-state exciton fine-structure should be the same for the size of nanocrystals used here. Further, energy transfer proceeds so rapidly as to inhibit spin manipulation. However, trap states for CdS excitons also exist, the influence of which is clearly seen in delayed PL where shelved excitons feed back into band edge states at times much longer than the exciton lifetime (discussed further in Sect. 3.6.2). We therefore manipulate the spin state of charge pairs shelved within the CdS, provided these maintain their spin identity while trapped. We do not directly manipulate those spins corresponding to the band edge exciton fine-structure. Spin resonance can then induce a conversion of mutual spin orientation for trapped carrier (electron–hole) pairs, converting them from "bright" to "dark" permutation symmetry. Once detrapping occurs, these weakly (exchange and magnetic dipole–) coupled spin pairs again become strongly coupled band edge exciton states where the mutual spin identity of the trapped carriers largely predetermines which excitonic fine-structure level becomes populated. Since the trap energies we concern ourselves with are, or are nearly, iso-energetic with the band gap, spin-scattering while moving in and out of trap-states is weak. This process of cycling carriers from band edge excitons to traps, changing trap state spin configuration, and then moving the carriers back to excitonic states is what generally allows spin-dependent PL in our structures: dark shelved carriers determine the population ratio for dark band edge excitons, which remain dark upon transfer to the CdSe core of the nanocrystal. It is important to note here that since PL is the observable in this scheme, at this time, a direct discrimination cannot be made between scenarios involving PL quenching due to an increase in trapping lifetime or quenching due to a direct transfer into a dark exciton [14, 21] state. In either case, the bright exciton population is diminished.

3.3 Spectrally Selected, Optically Detected Magnetic Resonance

Figure 3.1b summarizes the experimental approach (full details are provided in Sect. 3.6.1). A sample of tetrapods is illuminated by a continuous-wave laser and a homogeneous magnetic field splits the Zeeman sublevels. Transitions between these levels are induced coherently during the application of microwaves and, for optically active charge carriers, this process is witnessed as a transient perturbation in PL intensity with respect to the steady-state. A typical luminescence transient is illustrated in the figure: the microwave pulse should lead to luminescence *quenching* since optical excitation initially populates bright exciton states [21], but coherent spin mixing of intermediately shelved carriers leads to an overall increase of dark state exciton populations. After removal of the microwave field, the PL intensity returns slowly as shelved "dark" states undergo spin-lattice relaxation to form "bright" configurations which feed back into bright band edge excitons. This longer timescale process can result in an eventual *enhancement* over the steady-state background as long as the intersystem crossing rate is low relative to the rate of initial PL quenching [29]. The resonances of the composite CdSe/CdS material are surveyed in Fig. 3.2. In order to fully identify the material and spectral origin of observed resonant species, we compare separately CdSe quantum dots, CdS nanorods, and the full composite CdSe/CdS tetrapod heterostructures. The CdSe core emission spectrum is shown in Fig. 3.2a. The corresponding ODMR spectrum (panel (b)), where the differential PL is plotted as a function of magnetic field and time after the microwave pulse, shows only weak PL enhancement and no quenching, exhibiting broad inhomogeneity. We tuned the magnetic field over 1 T

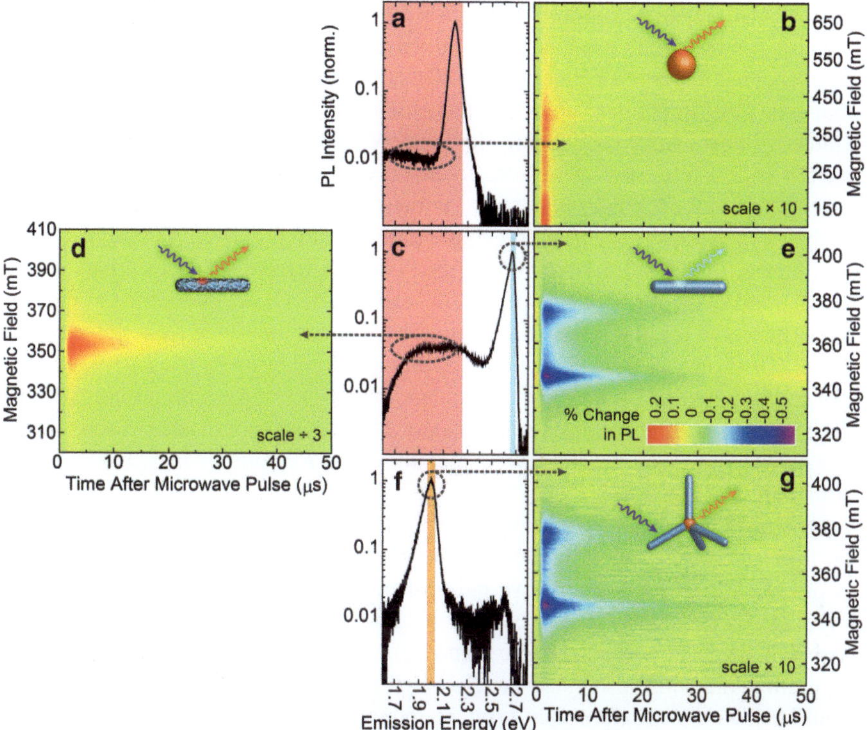

Fig. 3.2 Spectrally selected, spin-dependent transitions in semiconductor nanocrystals at 3.5 K under X-band (9.8 GHz) excitation. (**a**) Emission spectrum of bare CdSe nanocrystal quantum dots (tetrapod cores) and associated near-featureless transient ODMR spectrum (**b**) taken as a function of emission intensity (filter region marked in *red*) in dependence of magnetic field following a microwave pulse. (**c**) Emission spectrum of CdS nanorods. (**d**) Differential PL (enhancement) of CdS nanorod deep-trap level defect emission (marked *red* in panel (**c**)). (**e**) Differential PL (quenching) of the CdS band edge exciton emission (band labeled *blue* in panel (**c**)). (**f**) PL spectrum of CdSe/CdS nanocrystal tetrapods with associated transient ODMR spectrum detected in the CdSe emission (**g**), revealing the CdS spin species. The *Colored Bars* in panels (**a,c,f**) indicate the spectral region of the transmission filters used. The laser excitation energy is chosen to be just above the CdS nanorod band gap (~2.7 eV)

and found continuous PL enhancement over a range of 500 mT. The broad resonance is attributed to deep (below 2.1 eV) red-emitting highly spin-orbit coupled chemical defects of CdSe, and not to the band edge exciton [16, 17].

CdS nanorods are also known to emit at two energies; at ~2.667 eV (465 nm) due to the quantum-confined band edge exciton, and in a broad spectrum around 2.066 eV (600 nm) due to a deep-level chemical defect associated with a surface sulfur vacancy. The features are seen in the emission spectrum in panel (c). The ODMR transient mapping of the defect emission (selected by an emission filter) is illustrated in panel (d). A resonance is identified at 352 mT, corresponding to enhancement of defect PL, which decays over ~50 μs. In contrast, detection in the narrow exciton band (emission filter region marked blue in panel (c)) reveals distinct behavior (panel (e)): two resonances dominate, at 345 and 374 mT, corresponding to PL quenching under resonance. After ~30 µs, PL enhancement occurs. As discussed above, this transient interplay of PL quenching and enhancement is as expected for band edge trap states experiencing slow intersystem crossing, and intermixing with exciton states. In the following, we focus only on resonances associated with the exciton emission channel rather than luminescence of the defect, since the former likely relate to traps responsible for single-particle blinking[7–9]. As outlined below (and further in Sects. 3.6.3 and 3.6.4), the two band edge resonances arise due to a pair of weakly coupled spin-$\frac{1}{2}$ species, i.e. electron and hole. In contrast to the bare cores (panels (a,b)), the same CdSe emission spectrum measured from the tetrapods (panel (f)) shows ODMR characteristics that are dominated by the *CdS* band edge trap states (panel (g)). Here, spin-dependent transitions of the CdS are imparted on the core emission, enabling remote readout of CdS arm spin states. Such ODMR signals were only observed at low temperatures, their amplitude increasing steadily from 50 K down to 3.5 K.

To clarify the origin of spin-dependent transitions in the CdS exciton emission, we inspect the resonance dynamics. The nanorod ODMR spectrum in Fig. 3.3a, recorded 3.2 µs after a microwave pulse of 800 ns duration (i.e. a vertical slice of Fig. 3.2e), is accurately described by the sum of three Gaussian resonances. One peak is located at a characteristic Landé g-factor of $g = 2.0060(2)$ (blue arrow), suggesting that this resonance is related to a semifree charge [$g_{free-electron} \approx 2.002319$] with negligible spin-orbit coupling. The second distinct peak [black arrow, $g = 1.8486(2)$] is substantially shifted from the free-electron value, indicating that the carrier is localized in a trap with significant spin-orbit coupling. The third Gaussian is environmentally broadened (i.e. by hyperfine fields and a variation in effective spin-orbit coupling) and centered at $g = 1.9594(2)$ (grey arrow). Panel (b) plots the absolute differential PL against time after resonant microwave excitation for the black and blue peaks, revealing that the perturbed spin-state populations follow identical time dynamics during free-spin evolution. The decay of the pronounced initial quenching signal approximately follows a single exponential, indicating a dominant single spin-dependent transition rate. This transient is succeeded by a long-term PL enhancement, again dropping exponentially between 300 and 800 µs after microwave excitation. This form of decay, involving two primary exponential rates, is a clear signature of an electron–hole pair process [30]. Nearly identical resonance line shapes and dynamics are extracted for the tetrapods (panels (c,d)), confirming that spin information existing in the CdS nanorods can indeed be accessed via luminescence from the attached CdSe core. As seen in the comparison between the two sets of nanoparticles, line shapes and resonance center positions are expected to be subject to minor variations since both size and geometry of the particles affect quantum confinement and, therefore, the relative g-factors [14, 21, 28]. On average, the differential PL is ten times weaker for the tetrapods than for the nanorods, since light-harvesting of the CdS excitons inhibits trapping on metastable sites as required for this spin-resonant manipulation.

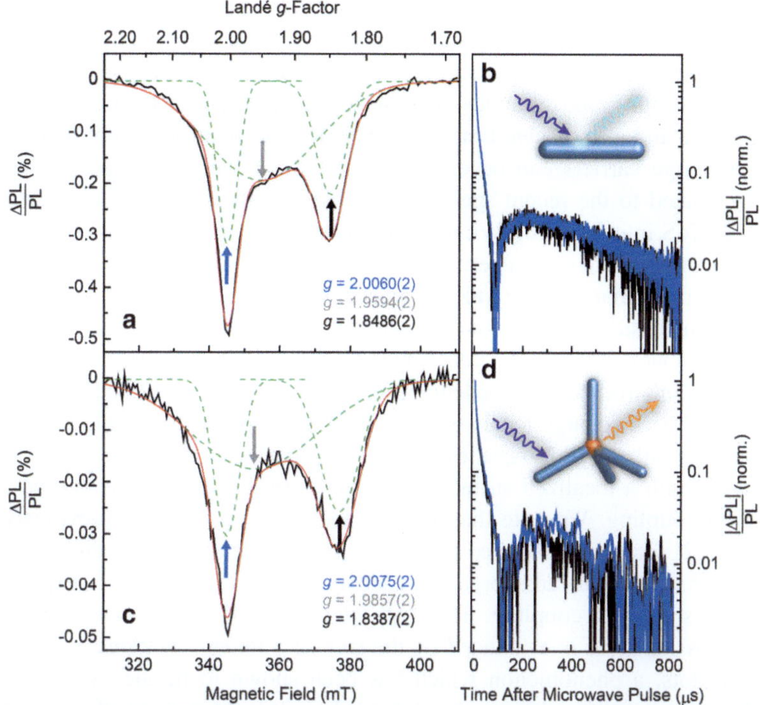

Fig. 3.3 Trapped charge correlation and remote readout of tetrapod arm states. Spectra at $3.2\,\mu s$ delay following an 800 ns microwave pulse detected in the band edge emission of CdS nanorods (**a**) and in the CdSe core emission of tetrapods (**c**). The spectra are accurately described by a superposition of three Gaussians. The temporal dynamics of the two dominant resonances (marked *blue* and *black*) are identical in (**b**) and (**d**), implying that the two spin-$\frac{1}{2}$ species are correlated. We assign these peaks to spin dynamics in a charge-separated state, with each charge carrier located on the surface of the nanocrystal. The charge at $g \approx 2.00$ represents a "semifree" carrier while the pair partner ($g \approx 1.84$) is situated on a site with greater spin-orbit interactions. The third broad Gaussian resonance follows different temporal dynamics and originates from an unrelated trapped species, located *within* the CdS where a large distribution in resonance frequencies exists. We tentatively assign the single broad resonance to a spin-dependent Auger-type process and the pair mechanism to the situation where both carriers are expelled from the bulk of the particle, generating surface charge which modulates fluorescence but leaves the particle neutral. Light harvesting in the tetrapods reduces the number of shelved excitons since carriers are rapidly removed from the CdS, leading to a tenfold reduction in signal strength

The resonances around $g \approx 2.00$ and $g \approx 1.84$ not only follow the same decay to equilibrium after spin-mixing but the spectral integrals also match (see Sect. 3.6.3). As discussed further in Sect. 3.6.4, this agreement is expected for a correlated spin-$\frac{1}{2}$ pair process; manipulation of either electron *or* hole spin has equal probability of modulating PL since the two charges couple by the same spin-dependent mechanism. An intriguing conclusion can be drawn from these observations: the band diagram of the tetrapods in Fig. 3.1 suggests that the hole should immediately localize in the CdSe core, although this is obviously not the case[31]. If this were the case, we would not observe identical resonance spectra and dynamics in nanorods and tetrapods. Instead, for the same spin-resonant manipulation of *both electron and hole* to occur in the tetrapods, both must be located within the CdS on the *same* nanoparticle and at the *same* time. The ODMR data therefore imply that trapping of both charge carriers can occur simultaneously at the band edge, a result that may be related to the recent spectroscopic identification of interfacial barriers at the CdSe/CdS interface [26]. Without significant modifications to the measurement technique or access to exact chemical information of at least one site, we are unable to assign a particular charge to these trap states since spin-resonance techniques are inherently insensitive to the sign of a charge.

Given the lack of spin-orbit coupling (a shift from the free-electron g-value) and only limited environmental broadening, we propose that the $g \approx 2.00$ peak originates from a "semifree" charge localized to the surface of the nanocrystal. The $g \approx 1.84$ resonance is only slightly broader than the $g \approx 2.00$ line, indicating that it is also associated with a localized surface site rather than the bulk, but is shifted due to spin-orbit coupling. We note that a resonance near $g \approx 2.00$ has previously been reported [32] for photogenerated holes in CdS [33], but this is also the expected g-value for charges localized to organic ligands[34] or matrix material[5] experiencing negligible spin-orbit coupling. This type of interaction with surface ligands is a distinct possibility as is evidenced by the lack of a phonon bottleneck in colloidal quantum dots, a phenomenon which has been shown to be mediated by carrier wavefunction overlap with organic ligands [35, 36]. At present, the information needed to precisely discriminate between these two chemical situations is not complete (see Sect. 3.6.7 for further discussion). The $g \approx 1.84$ feature is distinct from that found in ODMR of bulk CdS [37] ($g \approx 1.789$), although g-factors can shift significantly due to quantum size effects and geometry [14, 21, 28].

The third feature, the broad $g \approx 1.95$ peak marked grey in Fig. 3.3, only shows quenching and no enhancement, and decays faster than the narrow resonances, demonstrating that it arises from a distinct spin-dependent process (see Sect. 3.6.3). This feature vanishes in the tetrapods for excitation below the CdS band gap (see Sect. 3.6.6 for discussion). The broadening is likely induced by local strain or hyperfine fields, or by a superposition of multiple unresolved resonances. We propose that the resonance originates from a carrier trapped *within* the nanocrystal where a wide range of g-factors exists. This ODMR signal then likely arises due to spin-dependent Auger recombination [30] between the localized carrier and the quantum-confined band edge exciton within the particle, a process known to quench optical recombination [5, 7–9].

3.4 Coherence Measurements and Novel Effects

Rapid spin dephasing would normally be anticipated for a bulk-like crystal, given the significant spin-orbit coupling of the $g \approx 1.84$ resonance [15]. However, recording differential PL at each distinct resonance as a function of microwave pulse

Fig. 3.4 Spin dephasing of CdS trap states and coherent control of light-harvesting in tetrapods. (**a,b**) Rabi oscillations in the differential PL of the CdS nanorods as a function of microwave pulse length for the resonances around $g \approx 1.84$ and $g \approx 2.00$. The insets show the corresponding decay of spin coherence measured by performing Hahn spin echoes using a sequence of microwave pulses. (**c**) In the tetrapods, coherent spin information in the CdS is extracted remotely in the PL of the CdSe core, indicating the high degree of carrier localization since coherence information remains unperturbed upon change of environment (i.e. addition of the core to form the tetrapod heterostructure). This result also demonstrates the ability to coherently control the light-harvesting process

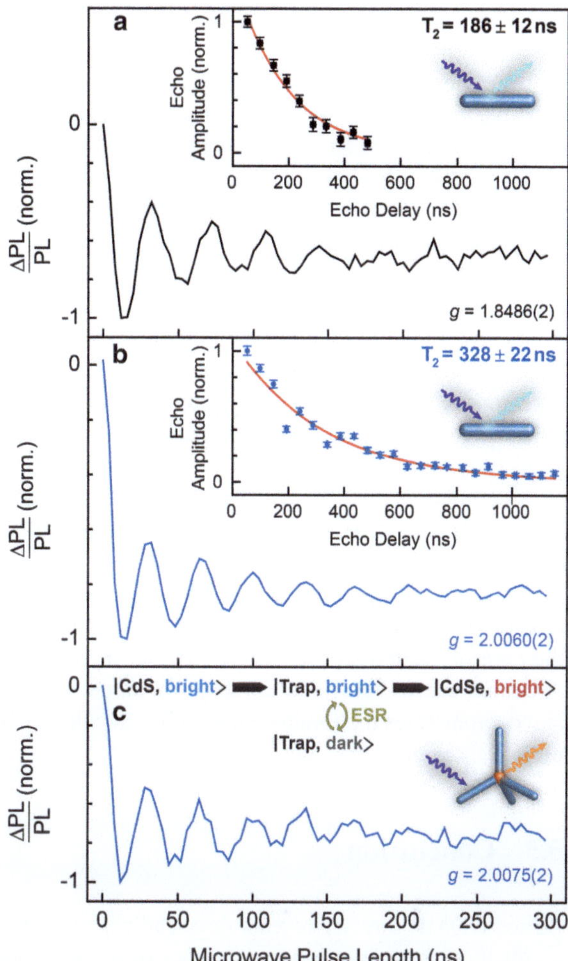

duration reveals Rabi flopping, as displayed in Fig. 3.4, a direct manifestation of spin-phase coherence. In this example, spins precess so that the shelved carrier pairs propagate *reversibly* between bright and dark mutual spin configurations. Such Rabi oscillations were recently reported for Mn-doped CdSe nanocrystals by conventional absorptive magnetic resonance[11], but are unprecedented for direct detection via intrinsic optical transitions of the semiconductor. The frequency components contained within this coherent oscillation provide additional information on the nature of these states; specifically on carrier spin-multiplicity and the existence of exchange and/or dipolar coupling. From this analysis (a detailed treatment is given in Sect. 3.6.4), it is found that both the $g \approx 2.00$ and $g \approx 1.84$ resonances describe carriers which carry spin-$\frac{1}{2}$. The mutual exchange and dipolar coupling experienced within the trapped pair is negligible.

Although the decay of the Rabi oscillation can provide a lower bound on the coherence lifetime for each of these carriers, more sophisticated resonant-pulse sequences can be used to unambiguously measure this value. We quantify the CdS spin-phase lifetime, T_2, by measuring Hahn spin echoes, the amplitude of differential PL change following rephasing of spins by a second microwave pulse (see full description in Sect. 3.6.5). Figure 3.4a, b (insets) exhibit exponential decay of the echo amplitude as a function of interpulse delay, τ, yielding $T_2 = 328 \pm 22$ ns for $g \approx 2.00$ and $T_2 = 186 \pm 12$ ns for $g \approx 1.84$. The coherence time of the $g \approx 1.95$ resonance is too short to be measured using our technique ($T_2 <$ several ns). Additional structure is seen on the echo decay of the $g \approx 2.00$ carrier due to hyperfine-field-induced electron spin-echo envelope modulation [11] (see Sect. 3.6.5). The pair partner of this quasi-free carrier, localized to a surface trap, experiences stronger spin-orbit coupling, lowering the g-factor and accelerating dephasing. Nevertheless, these T_2 values are unprecedented for nonmagnetic semi-conductor nanocrystals [11, 38].

It is notable that, as a consequence of these extraordinary coherence times, identical Rabi oscillations result under detection of CdS (nanorods, Fig. 3.4b) and CdSe emission (tetrapods, Fig. 3.4c). The experiments offer a qualitative assessment of the degree of trap localization. This must be significant since delocalized carriers would be expected to lose coherence by coupling to a new environment, such as the core of the tetrapods. The persistence of spin coherence over different system environments offers the possibility of remote readout of spin-phase information, and demonstrates the fundamental ability to coherently control light-harvesting [39] even in inorganic structures.

3.5 Conclusion

Pulsed ODMR directly reveals three radical species in CdS which control PL and are likely responsible for the two types of blinking observed in CdSe/CdS particles as distinguished by luminescence lifetime [8]: either both carriers are localized to the nanocrystal surface, leaving the particle neutral and thus preventing Auger recombination and a change in exciton lifetime ($g \approx 2.00$ and $g \approx 1.84$); or one carrier is trapped *within* the particle ($g \approx 1.95$), charging it so that Auger-type blinking with the associated fluorescence lifetime changes arises. This localization of carriers occurs in CdS, not CdSe. Surprisingly, shelved excitons do not ther-malize from CdS to CdSe, but remain in the CdS "shell" of the heterostructure nanoparticle [31]. The extraordinarily long spin quantum-phase coherence times of order $1\,\mu$s highlight the potential utility of even strongly spin-orbit-coupled nanoparticles for quantum information processing or quantum-enhanced sensing, such as magnetometry. In contrast to conventional inorganic quantum systems, such as electrostatically defined quantum dots, nanocrystals offer the possibility of creating spatially scalable quantum structures through bottom-up synthetic means [3] as demonstrated here by the spatially remote light-harvesting read-out of spin-phase information.

3.6 Supporting Information

3.6.1 Experimental Methods

The tetrapod nanocrystals consist of wurtzite CdS arms approximately 20 nm in length and 6 nm in diameter, grown onto four faces of zincblende CdSe cores of 4 nm diameter. Synthesis details are given in Ref. [1]. The same batch of CdSe cores which was used to seed tetrapod growth was also investigated alone for comparison, as shown in Fig. 3.2b of the main text. Each series of colloidal nanoparticles used in these measurements was first diluted into a toluene Zeonex (Zeon Chemicals L.P.) solution and then drop cast into a small Teflon bucket (2 × 3 mm). Upon solvent evaporation, a solid matrix was formed, which is both optically and paramagnetically inert, but contains the distributed nanoparticles. The sample was then suspended in a He flow cryostat containing a dielectric microwave resonator, generally kept at 3.5 K for all measurements, except for Rabi nutation experiments which were performed at 15 K. Optical access to the sample was made by extending a home-built fiber bundle through a cryostat port and into the resonator, resting at the mouth of the Teflon sample bucket. A cw Ar^+ laser, tuned to 457.9 nm (2.708 eV) and combined with a suitable filter to remove spontaneous emission (Semrock Maxline), was passed into a single fiber and used to excite the nanocrystal ensemble with 20 mW of power (intensity approximately 85 $\mu W/cm^{-2}$). The remainder of the fibers were used to collect PL, from which scattered laser light was filtered out with a 458 nm ultrasteep long-pass filter (Semrock RazorEdge). Specific emission bands for each nanoparticle ensemble were spectrally selected by choosing an appropriate filter set: the CdS nanorod and CdSe core deep-level defect emission were isolated with a 550 nm (2.254 eV) long-pass filter (ThorLabs); the CdS nanorod band edge emission was cut with a 460±2 nm (2.695±0.012 eV) narrow-band filter (ThorLabs); the tetrapod core emission was picked with a 620 ± 2 nm (2.000 ± 0.007 eV) narrow-band filter (ThorLabs).

The selected PL was focused onto a low-noise photodiode (Femto LCA-S-400-Si), whose signal was amplified with a Stanford Research Systems low-noise preamplifier (SR560). AC coupling of the input signal was used in order to apply gain to only the modulated contribution of the PL intensity. A 300 Hz high-pass frequency filter was also applied in order to help isolate the transient response of the ODMR signal from spurious electrical and optical modulations.

With sufficient gain applied, the resulting signal was passed into the fast digitizer of a Bruker SpecJet contained within an Elexsys E580 system, which correlates the timing of the microwave pulse sequence with the transient response. Programmable control over pulse routine timing, leveling of the external magnetic field, and signal acquisition was utilized to carry out the large number of measurements required for each data set. For example, the transient mappings displayed in Fig. 3.2c, d, f, h required an X-band (9.8 GHz) microwave pulse of 800 ns duration to be applied every 800 µs a total of 16,384 times. The transient responses of the individual measurements were added together before incrementing the external magnetic field B_0. For the high-resolution time transients shown in Fig. 3.3b, d, the microwave shot repetition rate was set to be greater than 2 ms, much longer than the full relaxation time to steady-state populations of carrier states under constant excitation of the material system. Rabi oscillations were obtained by monitoring the amplitude of the transient PL response as a function of microwave pulse length. The transit times for driving the system from optically dark to optically bright states served as useful timing information needed for constructing the π and $\pi/2$ pulses of the Hahn echo sequence. A full description of this conventional pulse sequence, as used in ODMR, is given below in Sect. 3.6.5.

3.6.2 Time-Resolved, Spectral Confirmation of Long-Lived Trap States

The existence of trap states lying very close to the band gap of our primary material system of interest, the CdS nanorods, can easily be confirmed by considering the luminescence decay characteristics following an optical excitation pulse. A sample similar to that used for the ODMR experiments is fabricated, consisting of nanorods suspended in a polystyrene block several microns thick. This sample is mounted, under vacuum, to the cold finger of a closed-cycle Helium cryostat, which cools to 21 K. A diode laser operating at 355 nm (3.493 eV) with nanosecond pulse length and variable repetition rate is used as an excitation source. PL spectra are monitored with a gated, intensified CCD (ICCD) camera mounted to a spectrometer, allowing us to record the decay of emission intensity as a function of gating time following optical excitation. The prompt PL is shown in Fig. 3.5. The dashed green line in panel (a) indicates the spectral position which is monitored as a function of time. As is seen in panel (b), the PL intensity drops off approximately following a power law over five orders of magnitude in time. The excitonic emission spectrum does not shift significantly over this time.

The accepted physical mechanism responsible for delaying emission in these nanoparticles for such long times is the temporary isolation of the optically excited charge carriers into their respective trap states [40], dramatically decreasing the amount of wavefunction overlap of the electron–hole pair, and therefore the likelihood of recombination. As the detrapping rate back into the band edge excitonic states depends exponentially on the trap energy, which in turn is distributed exponentially, a distribution of detrapping rates is observed across the nanoparticle ensemble, leading to the power law-like emission decay [40].

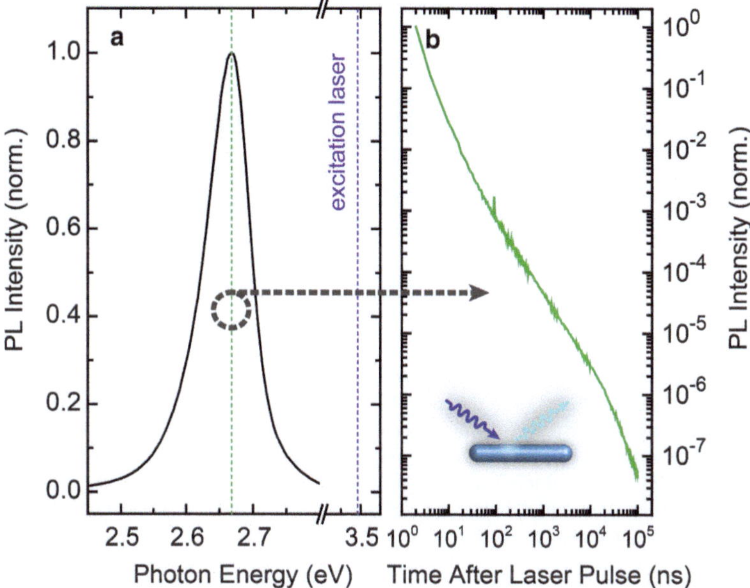

Fig. 3.5 Optical decay spectrum of CdS nanorods. (**a**) Prompt band edge PL spectrum of CdS nanorods at 21 K following excitation with a 355 nm (3.493 eV) laser pulse. (**b**) The emission peak is monitored as a function of delay time from excitation using a gated ICCD camera and spectrometer. The peak emission decay approximately follows a power law, revealing the presence of long-lived trap states, which are energetically close to the semiconductor band edge where the exciton forms

3.6.3   *Correlating Carrier-Pairs with Resonance Dynamics*

To determine which of the three resonances seen in the tetrapods and nanorods (Fig. 3.3) correspond to a coupled pair of carriers, the time dynamics of the resonances are considered. The correlation of the features in time dynamics in Fig. 3.3 is independent of temperature and laser power, although both of these parameters directly affect the transient response. The biexponential time dynamics shown in Fig. 3.3b, d are characteristic of a (electron–hole) pair process, which has been investigated extensively in the context of conjugated polymers [29].

While the observation of identical dynamics (Fig. 3.3) alone is sufficient to conclude that each of these paramagnetic centers belong to the same coupled system [29], further proof derives from a comparison of the areas of the two features. Since the area of each resonance represents the probability of inducing a spin transition which causes an optical activity, separately resonant carriers belonging to the same excitation (e.g. an electron and a hole in a pair) must exhibit equal probabilities for this process to occur. To aid in the analysis of comparing the equality of these probabilities, a fitting routine employing three Gaussians was utilized to study the transient spectra. For the CdS nanorod band edge emission,

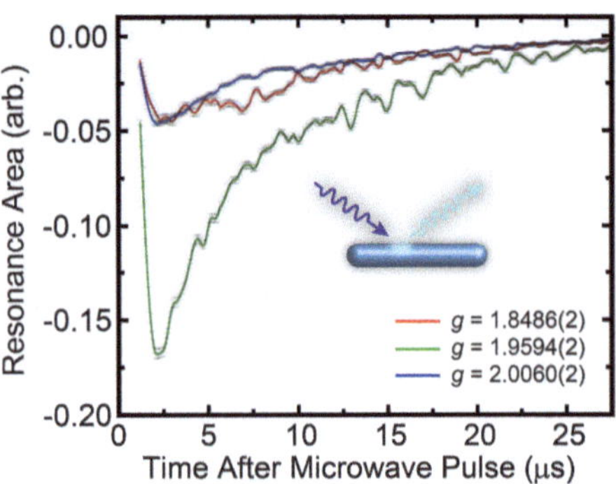

Fig. 3.6 The equality of resonance areas for coupled-pair trap states. Integrated resonances (areas) as a function of time obtained from the triple Gaussian fit applied to the ODMR mapping of CdS nanorod band edge emission. The equal resonance areas for the $g \approx 2.00$ and $g \approx 1.84$ sites denote the equal probabilities of inducing optical activity following a microwave-induced spin transition. Since the probabilities of inducing such a transition for each of the two sites are equal, it can be concluded that they represent a coupled pair of trap states (i.e. weakly bound electron–hole pair). The $g \approx 1.95$ state is clearly unrelated

the results of this analysis are shown in Fig. 3.6. A correlation of the $g \approx 2.00$ and $g \approx 1.84$ resonances is clear since the fitting routine finds comparable areas for the Gaussians representing these two features over a wide range of times. Additionally, we note that the central $g \approx 1.95$ resonance must represent a carrier state which is completely decoupled from the neighboring resonances since it displays marked differences in both probability (i.e. area of the resonance) and time dynamics.

A further point must be made about the disparity between the T_2 times given for each of these carriers in Fig. 3.4 in Sect. 3.4. The results of the Hahn echo experiment (outlined below in Sect. 3.6.5) on the $g \approx 1.84$ center of the CdS nanorods give a phase coherence time which is nearly half that of the $g \approx 2.00$ center, as would be expected for a carrier experiencing a larger degree of spin-orbit coupling. The inequality between T_2 times of the two (correlated) carriers reflects the unique chemical environments of each and does not conflict with the assignment of the two centers as representing a coupled pair. In fact, and although not measured explicitly, the only hard requirement imposed on the spin states of the pair is that each are characterized by identical T_1 times, which is inferred from the equal time dynamics of each resonance [29].

3.6.4 Spin and Carrier-Pair Interaction Information Obtained by Driven Rabi Oscillations

Determining spin identity is a crucial step in chemical fingerprinting as it can help to ultimately illuminate the chemical nature of a trap state for a specific carrier. For example, the complementary knowledge of spin multiplicity, resonance g-factor, and resonance structure can help to establish the exact symmetry of a paramagnetic site and, therefore, the exact environment of the localized carrier.

The spin identity of a paramagnetic center can be confirmed in a straight-forward manner by carrying out a Rabi nutation experiment since the carrier's spin quantum number is directly reflected in the frequency of oscillation between mutual spin configurations. For transitions between Zeeman-split m_s levels of the form $|S, m_s - 1\rangle \longrightarrow |S, m_s\rangle$, and neglecting any significant detuning from resonance, the Rabi frequency is determined by Astashkin and Schweiger[41] $\Omega_R = \sqrt{S(S+1) - m_s(m_s - 1)}$, where $\gamma = \frac{g\mu_B}{\hbar}$ is the gyromagnetic ratio for the center, μ_B is the Bohr magneton, $\hbar$ is Planck's constant, and B_1 is the microwave-induced magnetic field strength at the sample position within the microwave resonator. The g-factor is experimentally determined by the resonance center, but once the Rabi nutation has been recorded, a precise value for B_1 must be obtained in order to confirm the spin multiplicity of the trap site.

An additional material serving as a standard paramagnetic center with known g-factor and spin can be loaded into the microwave resonator alongside the material of interest; in this case, phosphorus-doped crystalline silicon (Si:P with a doping concentration of $[^{31}P] = 10^{16}$ cm^{-3}). The variations in microwave-induced magnetic field within the resonator volume that contains the combined sample are negligible over the few millimeters of sample breadth, allowing for the direct determination of B_1 fields experienced at the trap sites of the nanorods through recording of the Rabi frequency of ^{31}P centers in Si.

Aside from establishing the spin identity of the $g \approx 2.00$ and $g \approx 1.84$ sites, we are also interested in the type of mutual interactions experienced by the two carriers. Again, by scrutinizing the frequency components of the Rabi oscillations, general statements can be made as to the prevailing nature of intrapair coupling. The on-resonance spin-$\frac{1}{2}$ system precesses at a frequency of $\Omega_R = \gamma B_1$. As additional, nonnegligible interaction terms are introduced into the Hamiltonian describing the spin pair, further frequency components mix with Ω_R which directly correspond to specific forms of interactions. It has previously been shown [42] that increasing exchange interactions leads to frequency components of $2\gamma B_1$ appearing in the Rabi flopping signal, while an increase in dipolar interactions results in components of [43] $\sqrt{2}\gamma B_1$.

On the other hand, the same frequency components may arise not due to any particular pair interaction, but merely from spin transitions being stimulated within a particular spin manifold. For example, a $|\frac{3}{2}, \frac{-1}{2}\rangle \longrightarrow |\frac{3}{2}, \frac{1}{2}\rangle$ transition will produce a Rabi frequency component of $2\gamma B_1$, while a strongly exchange-coupled spin-$\frac{1}{2}$ system can do the same. This approach of attributing a systematic cause to a measured frequency component is made ambiguous if both the spin identity and the interaction type remain unresolved for the paramagnetic center. There is, however, one case where this ambiguity is easily resolved, which is for the spin-$\frac{1}{2}$ pair experiencing weak exchange and dipolar interactions. In this case, the only frequency component present in the Rabi nutation is γB_1, which is the case at hand.

Shown in Fig. 3.7 are Fourier transforms of the Rabi oscillations given in Fig. 3.4a ($g \approx 1.84$) and 3.4b ($g \approx 2.00$) of Sect. 3.4. We focus solely on data

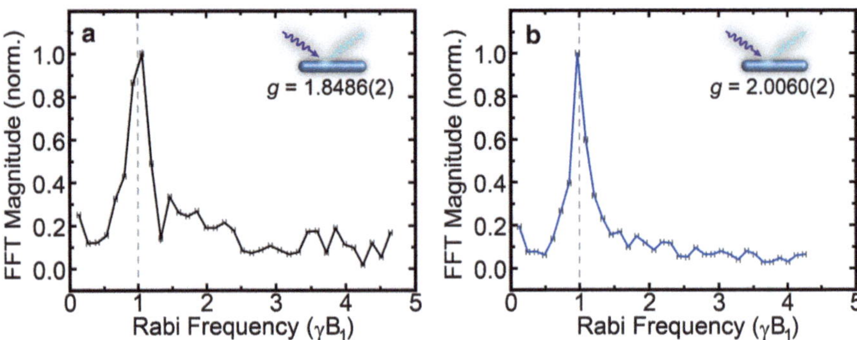

Fig. 3.7 Frequency components of Rabi oscillations show weak trapped-carrier coupling. Rabi oscillation frequency components are shown for both the $g \approx 1.84$ (**a**) and $g \approx 2.00$ (**b**) resonances of CdS nanorods. The frequency axis is scaled to γB_1, the Rabi frequency of a spin-$\frac{1}{2}$ paramagnetic center (marked by the *vertical dashed line*). There are no additional frequency components, demonstrating that each of these carriers is a spin-$\frac{1}{2}$ species and that the coupled pair of carriers experiences negligible exchange or dipolar coupling

obtained from the CdS nanorods to investigate the spin state and any possible interactions within the pair since the observed ODMR intensities of the nanorods are an order of magnitude larger than the same transitions seen in the tetrapods. The absence of any additional frequency components in the Fourier spectrum besides the γB_1 fundamental implies that this spin-dependent transition results from a pair of weakly-coupled spin-$\frac{1}{2}$ carriers, where both exchange and dipolar couplings are negligible. This weak coupling is not beyond expectations for such localized carriers since the distribution of trap sites over the nanoparticles should be random in space, leaving an average pair separation too large for either sufficient wavefunction overlap (exchange) or magnetic dipole-dipole interactions. Such weakly-bound precursor states, which ultimately feed into tightly bound band edge excitonic states, are common amongst a variety of material systems, such as hydrogenated amorphous silicon [43] and organic semiconductors [44], and can be manipulated through ESR in order to predetermine the permutation symmetry of final tightly bound states.

In this case, the carriers comprising this weakly bound precursor state are each spin-$\frac{1}{2}$, which means that they form mutual spin states that can be characterized as either singlet or triplet. This holds for the trapped carriers only, as the band edge excitonic states are well known to have a higher spin-multiplicity [21]. Therefore, upon detrapping, the singlet/triplet nature of the trapped pair will be projected upon the five individual spin states which make up the exciton fine structure. Since three of these states are bright and two are dark (i.e. spin allowed and forbidden optical transitions), changing the singlet/triplet nature of the trapped carriers will change the probability of moving back into a bright or dark state after detrapping occurs, thereby changing the overall exciton state populations.

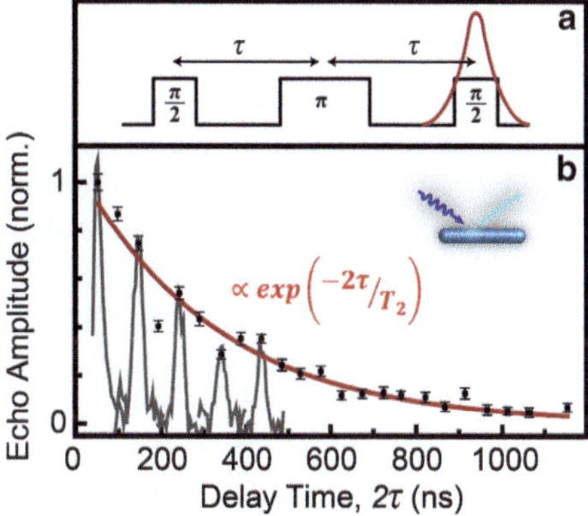

Fig. 3.8 Pulse timing diagram and demonstration of Hahn echo pulse sequence. (**a**) In the Hahn echo pulse sequence, a $\frac{\pi}{2}$-pulse projects the dominant initial population into a superposition of bright and dark states. Decoherence due to a distribution in local Larmor precession frequencies is reversed by application of a π-pulse after a delay time, τ. A second $\frac{\pi}{2}$-pulse projects the remaining superposition states (i.e. those which have not lost their spin phase information) back into a bright configuration. (**b**) Performing the measurement as a function of τ allows for the extraction of the characteristic spin dephasing time, T_2

3.6.5 *Measuring Spin Coherence and ESEEM with Optically Detected Hahn Echoes*

A lower limit on the spin dephasing time, T_2^*, of a paramagnetic center can be obtained by considering the amplitude decay of the Rabi oscillations. There are two primary mechanisms which artificially shorten coherence time in our system. One is due to the slight inhomogeneities in the oscillating magnetic field of the microwave radiation across the sample, ΔB_1, which leads to a distribution of Rabi frequencies, $\Delta\Omega_R$. Another is due to the distribution of local nuclear magnetic moments perturbing the static magnetic field, ΔB_0^{nuc}, experienced by the trapped carriers. This distribution leads to an additional detuning term in the Rabi frequency, further increasing $\Delta\Omega_R$. This spread in frequencies evolves the system towards incoherent transitions between the two spin configurations more quickly, but can be overcome by taking advantage of microwave pulse techniques to reveal the true dephasing time of the system, T_2. The Hahn echo pulse sequence is particularly appropriate [45]. Since the observable in ODMR is permutation symmetry (i.e. bright or dark mutual spin configuration) and not polarization as in traditional magnetic resonance, we use a slightly modified version of this classic technique. A simple $[\frac{\pi}{2} - \tau - \pi - \tau - \frac{\pi}{2}]$ pulse sequence is illustrated schematically in Fig. 3.8a, where a π-rotation denotes a complete reflection in permutation symmetry for the system and is determined by the precession time measured in a Rabi oscillation. The dynamics involved are straightforward. The first $\frac{\pi}{2}$-pulse places the initially bright spin population into a superposition of bright and dark states.

After a delay time, τ, in which the system dephases according to the distribution in Larmor frequencies arising from field inhomogeneities, a π-pulse is applied in order to reverse the Larmor precession of the system. This reversal effectively takes advantage of the time-reversal symmetry enforced by the long-time stability of the perturbing fields. The subsequent rephasing, or reversal in dephasing, takes place on a timescale equal to that of the initial delay, making the total dephasing time the system is subjected to 2τ. The second $\frac{\pi}{2}$-pulse is then applied in order to bring the remaining spin ensemble back to an observable state. By sweeping the $\frac{\pi}{2}$-pulse following the pulse sequence, a small change in the amplitude of transient response (i.e. the differential PL) is measured. This change is referred to as an echo, whose amplitude directly corresponds to the remainder of the initial population. By repeating this sequence and recording the echo as a function of 2τ, the loss of spin coherence is observed in the exponential decay of amplitude. Figure 3.8b illustrates this process by displaying some representative echoes using data for the CdS nanorod $g \approx 2.00$ center reproduced from the inset of Fig. 3.4b in the main text.

We note that the uncertainty of the quoted spin coherence time of the $g \approx 2.00$ resonance is higher than that of the $g \approx 1.84$ resonance, not due to improved signal-to-noise in the latter, but because of a complicated interference of the electron spin with nuclear magnetic moments. This substructure to the decay amplitude is apparent in Fig. 3.8b. It arises due to a phenomenon known as electron spin echo envelope modulation (ESEEM). Amplitude modulations of exponential decay of spin phase coherence would be expected to be present in the case of a finer structure splitting of the already Zeeman-split energy levels. These modulations do not prevent extraction of the decoherence time. With higher sensitivity, ESEEM should allow a precise chemical fingerprinting of the trap site in the future by providing information on local nuclear magnetic moments.

3.6.6 *Resonance Lineshapes vs. Optical Excitation Energy*

To probe the energetic distribution of trap centers, we studied the dependence of the CdSe/CdS tetrapod ODMR spectrum on excitation photon energy. This material system was chosen since excitation could be tuned from above the CdS arm band gap down to the absorption of the CdSe core while monitoring the resonance through the red-shifted PL of the CdSe core. Figure 3.9 shows the results of this excitation sequence. The broad, central resonance significantly decreases in amplitude at 488 nm (2.541 eV) excitation compared to 458 nm (2.708 eV) excitation, and disappears completely at 514 nm (2.412 eV). The coupled-pair resonances ($g \approx 2.00$ and $g \approx 1.84$) remain intact, although some broadening is observed with decreasing excitation energy. This observation implies that the species represented by the central resonance has a unique relationship to the delocalized band edge states of

Fig. 3.9 The dependence of the tetrapod ODMR spectrum on excitation energy. As excitation energy is decreased, the central resonance disappears, suggesting a close relationship to band edge states. The coupled-pair resonances remain, shifting slightly in center position and broadening

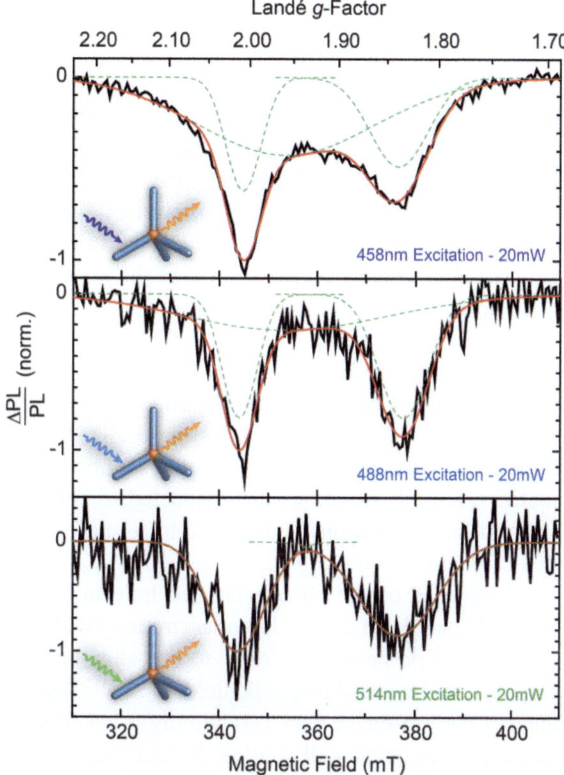

CdS, as compared to the $g \approx 2.00$ and $g \approx 1.84$ coupled-pair species, which exist over a much broader distribution of excitation energies. As commented on in the main text, this observation suggests that the narrow pair species both correspond to CdS surface states which can be populated even by direct excitation of the CdSe core slightly below the CdS band edge. Due to the presence of lattice strain at the heterojunction interface [46], carriers can still become trapped in the CdS even if the excitation energy lies below the band gap of the CdS nanorod. The involvement of lattice strain may explain the slight broadening of the $g \approx 2.00$ and $g \approx 1.84$ coupled-pair resonances with decreasing excitation energy. These resonances are comparatively narrow, suggesting that they correspond to discrete atomic sites such as surface defects, organic ligands, or the surrounding organic matrix. In contrast, the broad resonance can only be excited when the CdS is pumped above the band edge, suggesting that this species originates from bulk delocalized states in the CdS with substantial disorder broadening due to a wide range of chemical environments probed.

3.6.7 Possible Origins of the $g = 2$ Resonance

The resonance present in the CdS nanorods which is most closely aligned with the free-electron g-value is that at $g = 2.0060(2)$. Although the same center is observed through the CdSe core emission of the tetrapod structure, since the deep-level chemical defect of the CdS arm also emits at this energy (Fig. 3.2c), the exact resonance position is likely perturbed due to convolution with the resonance structure of the defect (Fig. 3.2d). To avoid this convolution, we rely on the CdS nanorod data to most accurately assess the features of the $g = 2.0060(2)$ resonance, since in this case, the excitonic and defect states are spectrally well separated. In describing the nature of the $g \approx 2.00$ resonance, there are two possible models which are supported in the literature. One involves a photogenerated hole becoming trapped at the CdS surface in an undetermined chemical position [33]. Another possibility is that a charge becomes localized to an incorrectly-bonded surface-ligand site [34], or ejected from the nanoparticle into the surrounding organic matrix [5]. Both situations are suspected to constitute a type of charge trap [34]. Each of these situations is expected to result in a resonance position very close to that of the free-electron g-factor ($g_{free-electron} \approx 2.002319$).

In the case of the photogenerated hole in CdS, Ref. [33] reported such a site which displayed an axial g-factor asymmetry with $g_\parallel = 2.035$ and $g_\perp = 2.005$, where parallel and perpendicular refer to the alignment of principal g-factor axes with respect to the external magnetic field, $\vec{B}_0$. For a disordered ensemble of nanocrystals, each of these g-factor axes is randomly oriented with respect to $\vec{B}_0$ and so the spin resonance spectrum will display distinct peaks for each principal g-value, as well as a continuum of peaks between these values representing the linear combination of projections. Such a lineshape is referred to as an anisotropic powder pattern. In general, $g_\perp$ results in a higher degree of spin-polarization due to the larger number of axis-normal orientations expressed in the random distribution. This effect would give maximal resonant change in photoluminescence at $g_\perp$, as compared to $g_\parallel$, which is very near to the situation we observe here.

In considering the second case, that of the charge localized to some organic material (either ligands or matrix), we also find good agreement between our measurements and the expected characteristics for such a material. Due to the extremely low levels of spin-orbit coupling, the g-factor of organic materials is found to be quite close to the free-electron value. Consequently, for a charge which is localized to a surface passivating organic ligand, a resonance very close to $g \approx 2.00$ would be expected. In addition to the resonance position, the value for the T_2 coherence time determined is much longer than those reported for similar inorganic quantum dots [38], yet is of the order of that measured in organic semiconductor systems [38]. Since neither the organic ligands nor the matrix are π-conjugated, it is not presently clear how their chemical structures would support charging, although defect centers respective to these materials are conceivable.

Discrimination between these two models remains difficult at this time without additional information. The level of inhomogeneous broadening and the overlap of the $g \approx 1.95$ resonance presently prevent us from resolving in detail any possible anisotropic features of this resonance. In the previous report of photogenerated holes in CdS[33], detailed information regarding the relative amplitude difference and line width difference between the $g_\perp$ and $g_\parallel$ spectral positions is lacking. We therefore resort to using a single Gaussian line profile in order to represent this resonance as a type of first-order approximation. More parameter information is necessary (line widths and peak intensity ratios) to faithfully make use of a powder pattern fitting function in comparing these two models.

Resolving the ambiguity of chemical assignment for this resonance site could be carried out in at least two ways. One is in using an electron-spin-echo (ESE) detection scheme in order to map out the resonance structure. This technique allows one to independently measure the resonance structure of two overlapping species which have differing coherence times. The ability to separate out the overlapping $g \approx 1.95$ resonance may result in finer resolution of the $g \approx 2.00$ feature details and therefore resolve the issue of line shape anisotropy. A second method of resolving this issue is through taking advantage of the slight amplitude modulation which is likely to be present in the Hahn echo decay of this center, known as ESEEM (described above). By measuring this modulation with higher resolution, both in amplitude and echo delay spacing, the frequency components involved should allow discrimination between the trapped spin interacting with either a local ^{1}H, ^{111}Cd, or ^{113}Cd nuclear magnetic moment. Such a measurement would give a direct chemical fingerprint of the trap site position.

References

1. Talapin, D.V., Nelson, J.H., Shevchenko, E.V., Aloni, S., Sadtler, B., Alivisatos, A.P.: Seeded growth of highly luminescent CdSe/CdS nanoheterostructures with rod and tetrapod morphologies. Nano Lett. **7**(10), 2951–2959 (2007)
2. Ip, A.H., Thon, S.M., Hoogland, S., Voznyy, O., Zhitomirsky, D., Debnath, R., Levina, L., Rollny, L.R., Carey, G.H., Fischer, A., Kemp, K.W., Kramer, I.J., Ning, Z., Labelle, A.J., Chou, K.W., Amassian, A., Sargent, E.H.: Hybrid passivated colloidal quantum dot solids. Nat. Nanotechnol. **7**(9), 577–582 (2012)
3. Milliron, D.J., Hughes, S.M., Cui, Y., Manna, L., Li, J., Wang, L.-W., Alivisatos, A.P.: Colloidal nanocrystal heterostructures with linear and branched topology. Nature **430**, 190–195 (2004)
4. Jones, M., Lo, S.S., Scholes, G.D.: Quantitative modeling of the role of surface traps in CdSe/CdS/ZnS nanocrystal photoluminescence decay dynamics. Proc. Natl. Acad. Sci. **106**(9), 3011–3016 (2009)
5. Efros, A.L., Rosen, M.: Random telegraph signal in the photoluminescence intensity of a single quantum dot. Phys. Rev. Lett. **78**(6), 1110–1113 (1997)
6. Schlegel, G., Bohnenberger, J., Potapova, I., Mews, A.: Fluorescence decay time of single semiconductor nanocrystals. Phys. Rev. Lett. **88**(13), 137401 (2002)
7. Galland, C., Ghosh, Y., Steinbrück, A., Sykora, M., Hollingsworth, J.a., Klimov, V.I., Htoon, H.: Two types of luminescence blinking revealed by spectroelectrochemistry of single quantum dots. Nature **479**(7372), 203–207 (2011)
8. Rosen, S., Schwartz, O., Oron, D.: Transient fluorescence of the off state in blinking CdSe/CdS/ZnS semiconductor nanocrystals is not governed by Auger recombination. Phys. Rev. Lett. **104**(15), 157404 (2010)

9. Zhao, J., Nair, G., Fisher, B.R., Bawendi, M.G.: Challenge to the charging model of semiconductor-nanocrystal fluorescence intermittency from off-state quantum yields and multiexciton blinking. Phys. Rev. Lett. **104**(15), 157403 (2010)
10. Kraus, R.M., Lagoudakis, P.G., Rogach, A.L., Talapin, D.V., Weller, H., Lupton, J.M., Feldmann, J.: Room-temperature exciton storage in elongated semiconductor nanocrystals. Phys. Rev. Lett. **98**(1), 017401 (2007)
11. Ochsenbein, S.T., Gamelin, D.R.: Quantum oscillations in magnetically doped colloidal nanocrystals. Nat. Nanotechnol. **6**(2), 112–115 (2011)
12. Crooker, S., Awschalom, D., Baumberg, J., Flack, F., Samarth, N.: Optical spin resonance and transverse spin relaxation in magnetic semiconductor quantum wells. Phys. Rev. B **56**(12), 7574–7588 (1997)
13. Nirmal, M., Norris, D., Kuno, M., Bawendi, M., Efros, A., Rosen, M.: Observation of the "dark exciton" in CdSe quantum dots. Phys. Rev. Lett. **75**(20), 3728–3731 (1995)
14. Gupta, J.A., Awschalom, D.D., Efros, A.L., Rodina, A.V.: Spin dynamics in semiconductor nanocrystals. Phys. Rev. B **66**(12), 125307 (2002)
15. Scholes, G.D., Kim, J., Wong, C.Y.: Exciton spin relaxation in quantum dots measured using ultrafast transient polarization grating spectroscopy. Phys. Rev. B **73**(19), 195325 (2006)
16. Glozman, A., Lifshitz, E.: Optically detected spin and spin-orbit resonance studies of CdSe/CdS core-shell nanocrystals. Mater. Sci. Eng. C **15**(1–2), 17–19 (2001)
17. Gokhberg, K., Glozman, A., Lifshitz, E., Maniv, T., Schlamp, M.C., Alivisatos, P.: Electron (hole) paramagnetic resonance of spherical CdSe nanocrystals. J. Chem. Phys. **117**(6), 2909–2913 (2002)
18. Lifshitz, E., Glozman, A., Litvin, I.D., Porteanu, H.: Optically detected magnetic resonance studies of the surface/interface properties of II–VI semiconductor quantum dots. J. Phys. Chem. B **104**(45), 10449–10461 (2000)
19. Fradkin, L., Langof, L., Lifshitz, E., Gaponik, N., Rogach, A., Eychmüller, A., Weller, H., Micic, O.I., Nozik, A.J.: A direct measurement of g-factors in II–VI and III–V core-shell nanocrystals. Phys. E **26**(1–4), 9–13 (2005)
20. Merkulov, I.A., Efros, A.L., Rosen, M.: Electron spin relaxation by nuclei in semiconductor quantum dots. Phys. Rev. B **65**(20), 205309 (2002)
21. Efros, A., Rosen, M., Kuno, M., Nirmal, M., Norris, D., Bawendi, M.: Band-edge exciton in quantum dots of semiconductors with a degenerate valence band: dark and bright exciton states. Phys. Rev. B **54**(7), 4843–4856 (1996)
22. Liu, S., Borys, N., Huang, J., Talapin, D., Lupton, J.: Exciton storage in CdSe/CdS tetrapod semiconductor nanocrystals: electric field effects on exciton and multiexciton states. Phys. Rev. B **86**(4), 045303 (2012)
23. Rocke, C., Zimmermann, S., Wixforth, A., Kotthaus, J., Böhm, G., Weimann, G.: Acoustically driven storage of light in a quantum well. Phys. Rev. Lett. **78**(21), 4099–4102 (1997)
24. Frantsuzov, P.A., Marcus, R.: Explanation of quantum dot blinking without the long-lived trap hypothesis. Phys. Rev. B **72**(15), 155321 (2005)
25. Gómez-Campos, F.M., Califano, M.: Hole surface trapping in CdSe nanocrystals: dynamics, rate fluctuations, and implications for blinking. Nano Lett. **12**(9), 4508–4517 (2012)
26. Borys, N.J., Walter, M.J., Huang, J., Talapin, D.V., Lupton, J.M.: The role of particle morphology in interfacial energy transfer in CdSe/CdS heterostructure nanocrystals. Science **330**, 1371–1374 (2010)
27. Müller, J., Lupton, J., Rogach, A., Feldmann, J., Talapin, D., Weller, H.: Monitoring surface charge migration in the spectral dynamics of single CdSe/CdS nanodot/nanorod heterostructures. Phys. Rev. B **72**(20), 205339 (2005)
28. Horodyská, P., Němec, P., Sprinzl, D., Malý, P., Gladilin, V.N., Devreese, J.T.: Exciton spin dynamics in spherical CdS quantum dots. Phys. Rev. B **81**(4), 045301 (2010)
29. McCamey, D.R., Lee, S.-Y., Paik, S.-Y., Lupton, J.M., Boehme, C.: Spin-dependent dynamics of polaron pairs in organic semiconductors. Phys. Rev. B **82**(12), 125206 (2010)
30. Baker, W., McCamey, D., van Schooten, K., Lupton, J., Boehme, C.: Differentiation between polaron-pair and triplet-exciton polaron spin-dependent mechanisms in organic light-emitting diodes by coherent spin beating. Phys. Rev. B **84**(16), 165205 (2011)

31. Mauser, C., Da Como, E., Baldauf, J., Rogach, A.L., Huang, J., Talapin, D.V., Feldmann, J.: Spatio-temporal dynamics of coupled electrons and holes in nanosize CdSe-CdS semiconductor tetrapods. Phys. Rev. B **82**(8), 081306 (2010)
32. Keeble, D.J., Thomsen, E.A., Stavrinadis, A., Samuel, I.D.W., Smith, J.M., Watt, A.A.R.: Paramagnetic point defects and charge carriers in PbS and CdS nanocrystal polymer composites. J. Phys. Chem. C **113**(40), 17306–17312 (2009)
33. Nakaoka, Y., Nosaka, Y.: Electron spin resonance study of radicals produced on irradiated CdS powder. J. Phys. Chem. **99**(24), 9893–9897 (1995)
34. Voznyy, O.: Mobile surface traps in CdSe nanocrystals with carboxylic acid ligands. J. Phys. Chem. C **115**, 15927–15932 (2011)
35. Guyot-Sionnest, P., Wehrenberg, B., Yu, D.: Intraband relaxation in CdSe nanocrystals and the strong influence of the surface ligands. J. Chem. Phys. **123**(7), 074709 (2005)
36. Cooney, R.R., Sewall, S.L., Anderson, K.E.H., Dias, E.A., Kambhampati, P.: Breaking the phonon bottleneck for holes in semiconductor quantum dots. Phys. Rev. Lett. **98**(17), 177403 (2007)
37. Brunwin, R.F., Cavenett, B.C., Davies, J.J., Nicholls, J.E.: Optically detected donor magnetic resonance in CdS. Solid State Commun. **18**(9–10), 1283–1285 (1976)
38. Whitaker, K.M., Ochsenbein, S.T., Smith, A.L., Echodu, D.C., Robinson, B.H., Gamelin, D.R.: Hyperfine coupling in colloidal n-type ZnO quantum dots: effects on electron spin relaxation. J. Phys. Chem. C **114**(34), 14467–14472 (2010)
39. Herek, J.L., Wohlleben, W., Cogdell, R.J., Zeidler, D., Motzkus, M.: Quantum control of energy flow in light harvesting. Nature **417**, 533–535 (2002)
40. van Driel, A.F., Nikolaev, I.S., Vergeer, P., Lodahl, P., Vanmaekelbergh, D., Vos, W.L.: Statistical analysis of time-resolved emission from ensembles of semiconductor quantum dots: interpretation of exponential decay models. Phys. Rev. B **75**(3), 035329 (2007)
41. Astashkin, A.V., Schweiger, A.: Electron-spin transient nutation: a new approach to simplify the interpretation of ESR spectra. Chem. Phys. Lett. **174**(6), 595–602 (1990)
42. Gliesche, A., Michel, C., Rajevac, V., Lips, K., Baranovskii, S.D., Gebhard, F., Boehme, C.: Effect of exchange coupling on coherently controlled spin-dependent transition rates. Phys. Rev. B **77**(24), 245206 (2008)
43. Herring, T.W., Lee, S.Y., McCamey, D.R., Taylor, P.C., Lips, K., Hu, J., Zhu, F., Madan, A. Boehme, C.: Experimental discrimination of geminate and non-geminate recombination in a-Si:H. Phys. Rev. B **79**(19), 195205 (2009)
44. Lee, S.-Y., Paik, S.-Y., McCamey, D.R., Yu, J., Burn, P.L., Lupton, J.M., Boehme, C.: Tuning hyperfine fields in conjugated polymers for coherent organic spintronics. J. Am. Chem. Soc. **133**(7), 2019–2021 (2011)
45. Hahn, E.L.: Spin echoes. Phys. Rev. **297**(1946), 580–594 (1950)
46. Smith, A.M., Nie, S.: Semiconductor nanocrystals: structure, properties, and band gap engineering. Acc. Chem. Res. **43**(2), 190–200 (2010)

Chapter 4
Towards Chemical Fingerprinting of Deep-Level Defect Sites in CdS Nanocrystals by Optically Detected Spin Coherence

4.1 Chapter Synopsis

Carrier trapping in colloidal nanocrystals represents a major energy loss mechanism for excitonic states crucial to devices, yet surprisingly little is known about the chemical nature of these trap centers or the types of interactions that charges experience in them. Here, we use a pulsed microwave optically detected magnetic resonance (pODMR) technique in order to probe the interaction pathways existing between shallow band edge trap states and the deep-level emissive chemical defect states responsible for the broad, low energy emission common to CdS nanocrystals. Due to the longer spin-coherence lifetimes (T_2) of these states, Rabi flopping in the differential luminescence under resonance provides access to information regarding coupling types of shallow-trapped electron-hole pairs, both isolated species and those in proximity to the emissive defect. Corresponding Hahn spin-echo measurements expose an extraordinary long spin coherence time for colloidal nanocrystals ($T_2 \approx 1.6\,\mu s$), which allows observation of local environmental interactions through electron spin-echo envelop modulation (ESEEM). Such an effect provides future opportunities for gaining the detailed chemical and structural information needed in order to eliminate energy loss mechanisms during the synthetic process.

4.2 Introduction

Substantial advances in the fundamental understanding of electronic states characterizing colloidal nanocrystals have been made in recent years, which have facilitated the development of novel and exciting device concepts based on this unique material system. These proposed technologies range from next-but-one-generation photovoltaics [1], over multicolor lasers [2], to inkjet-printed LED

K. van Schooten, *Optically Active Charge Traps and Chemical Defects in Semiconducting Nanocrystals Probed by Pulsed Optically Detected Magnetic Resonance*, Springer Theses, DOI 10.1007/978-3-319-00590-4_4, © Springer International Publishing Switzerland 2013

displays on flexible substrates [3]. Plaguing the further development of such devices has been the existence of charge trap states and chemical defects [4, 5] which naturally arise during the traditional synthesis process, but provide strong alternative decay pathways [6] for the desired excitonic and multi-excitonic excited states.

The origin of these detrimental states has been attributed to surface dangling bonds due to ligand loss [7], incorrectly bonded passivation ligands [8], as well as crystalline defects such as vacancies and adatoms [9, 10]. The effects of such states are readily detected through observables of conventional optical probes, such as by monitoring single particle luminescence intermittency [11–13] (i.e. "blinking") and the broad, sub-band gap chemical defect emission [9, 10, 14]. In fact, several fairly complex models have been formulated in order to describe the complexity witnessed in photoluminescence (PL) blinking [15, 16] and decay dynamics [17–19], which generally depend on certain assumptions about the population and decay pathways of both band edge and trap/defect states and their respective interactions. Even though these models go to great lengths in order to describe the complex dynamics observed experimentally, they are often not detailed enough, since they normally do not account for the existence of multiple trap species and both electron and hole traps, which has recently been confirmed for at least one type of nanocrystal (see Chap. 3). In general, very little is actually known about the chemical nature of these trap states or the types of interactions they experience since they are difficult to address directly using optical techniques alone. Spin resonance methods, on the other hand, are uniquely suited as a probe for such states and have historically proven to be a powerful tool in elucidating the chemical and electronic nature of charge traps and structural defects in a wide range of semiconductor systems [20–22].

Here, we use pulsed optically detected magnetic resonance (pODMR) in order to directly probe trapped carriers which are associated with both band edge as well as deep-level chemical defect emission in wurtzite CdS nanorods. It is well known that band edge excitons can be "shelved" in band edge trap states, leading to delayed PL at times much longer than the exciton lifetime [18]. We observe that these charge traps, which shelve the primary exciton [23], are capable of interacting with both band edge excitonic states (leading to emission from the quantum-confined exciton) as well as with the emissive deep-level chemical defect. The latter case leads to a modification of the spin resonance properties of the band edge trap states. We explore the trap states which are more directly associated with the chemical defect emission process, demonstrating that these are spatially highly localized with substantial dipolar coupling between carrier spins, as is most clearly manifested in the appearance of a half-field resonance. These states can be utilized as an environmental probe through ESEEM, which becomes possible due to the extraordinarily long coherence time of the state ($T_2 \approx 1.6\,\mu s$) at 3.5 K.

4.3 Photoluminescence Decay Dynamics Indicating Long Trapping Lifetimes

The pODMR spin-resonance technique is limited by a carrier's lifetime within a particular state relative to the timescale of spin-mixing induced by a resonant microwave pulse (~10 ns). In this case, the trapping lifetime must be long and trapped carriers must directly feed one of the two primary emission channels, our observables for this material system: that is, the excitonic band edge emission at 464 nm, and the deep-level chemical defect emission at ~635 nm [9]. Optical investigations of this defect emission in CdS nanocrystals, both with [10] and without [9] surface charge modification, have concluded that the emissive center is strongly related to surface S^{2-} and Cd^{2+} ion vacancies which act as electron and hole traps, respectively, and likely form a defect cluster acting as a color center [9]. To confirm that the trap states do indeed have sufficiently long lifetimes, we consider the PL decay dynamics of an ensemble of CdS nanorods following an optical excitation pulse. A sample of these nanocrystals is suspended in a polystyrene block several microns thick, which is mounted to the cold finger of a closed-cycle He cryostat operating at 21 K. Pulsed optical excitation is achieved with a 355 nm diode laser and the resultant PL spectrum is captured as a function of time with a gated, intensified CCD (ICCD) camera which is mounted to a spectrometer. Figure 4.1a shows prompt (2 ns integration window) and delayed (10 μs delay, 0.1 μs integration window) emission spectra, revealing the two distinct emissive species: the narrow blue exciton band which dominates the prompt emission; and the broad red defect band which appears at longer times. Recording the spectral decay following subnanosecond excitation allows confirmation of the presence of long-lived trap states [17–19] feeding these two emission channels, which have distinct lifetimes as seen in Fig. 4.1b. Besides the requirement for long lifetimes of suitable carrier states, pODMR additionally requires sufficient lifetimes of the spin state (T_1 > a few ns), which is also satisfied for several of the trap states existing in CdS.

4.4 Experimental Methods

For all pODMR measurements, a sample similar to the above is fabricated, with CdS nanorods dispersed within an optically inert, diamagnetic matrix. The sample is then held at cryogenic temperatures within a He flow cryostat and a low-Q dielectric resonator of a Bruker E580 pulsed electron spin resonance (ESR) spectrometer. Optical excitation is carried out with an Ar^+ laser tuned to 457.9 nm, which is almost resonant with the band gap of CdS nanorods of this diameter (6 nm). As band edge excitonic states are generated optically, there is some probability for them to become localized to the shallow trap states which are nearly iso-energetic with the band gap, or to lose energy and become associated with a deeper-lying chemical defect. Once charge carriers become trapped, they may either detrap and return to

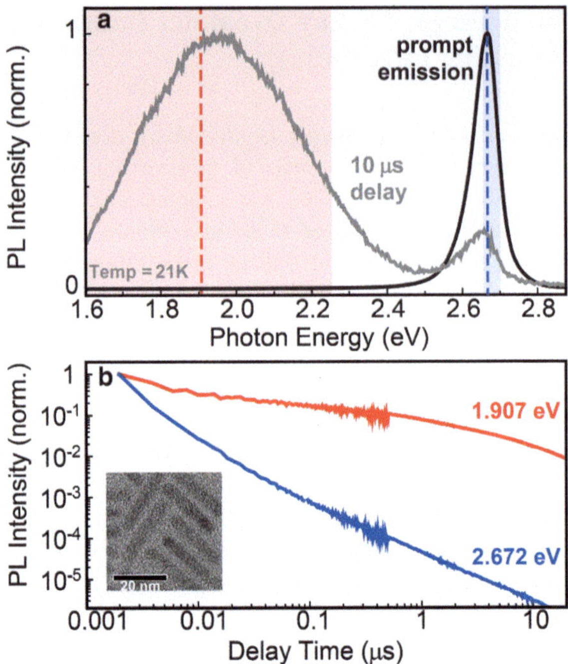

Fig. 4.1 Time-resolved luminescence of CdS nanorods exhibiting dual emission fed by long-lived trap states. (**a**) Prompt (0–2 ns integration window) and 10 µs delayed (10–10.1 µs integration window) optical emission spectra following 0.7 ns pulsed excitation from a 355 nm diode laser. A time-gated ICCD coupled to a spectrometer is used in acquisition. Prompt emission is dominated by band edge exciton emission, whereas the broad, *red* emission channel is attributed to deep-level chemical defect states. The *light blue* and *red* regions denote the spectral bandwidth of collection filters used for luminescence channel isolation in the ODMR experiments. *Vertical blue* and *red dashed lines* mark the spectral positions used to demonstrate the existence of long-lived trap states feeding each emission channel, as is evident by the long power-law-like PL decays given in panel (**b**). The integration window used for each step in time delay was: 2 ns for 0–510 ns; 20 ns for 0.51–2.0 µs; 100 ns for 2.0–10 µs; 1.0 µs for 10–20 µs. The *inset* shows a transmission electron micrograph illustrating the high quality of CdS nanorods

excitonic states, feed into nearby deep-level chemical defect states, interact with subsequent optical excitations through an Auger process, or simply thermalize to the ground state. These processes are spin dependent, so that spin manipulation in pODMR becomes possible. In the present work we focus on the spin-dependent emission characteristics of the deep-level defect by using coherent modulations of the emission as a probe of the intermediate states involved in populating the defect and their respective environments.

The optically detected spin resonance process arises by first Zeeman splitting the spin states associated with optically active carrier pairs by applying an external magnetic field (~0.3 T) to the sample and then matching that energy splitting with a pulsed microwave field (~9.8 GHz; with the frequency held constant in all measurements). The resonant change in mutual spin identity for the carrier pair is reflected in PL intensity as a resultant enhancement or quenching of the emission channel [24], as observed by acquisition from a low-noise photodiode (FEMTO LCA-S-400K-SI). In principle, individual trapping species can be fully resolved since the resonance condition for each type is unique and corresponds to the Landé g-factor of that state. In order to differentiate between trap species which play a role in separate emission processes, pODMR is performed on each of the two CdS nanorod emission channels by using optical selection filters (spectral bandwidths given in Fig. 4.1a). This general process of excitation, charge trapping or localization on a defect, and ESR of optically active carriers is schematically depicted in Fig. 4.2a. The results of this spectrally resolved pODMR can be seen for each emission channel in Fig. 4.2b, d, where the resonant change in PL intensity, plotted on a color scale, is shown as a function of magnetic field strength and time following the microwave pulse. The temporally integrated magnetic field dependence of differential PL yields the resonance spectra in Fig. 4.2c, e.

4.5   Optically Detected Magnetic Resonance vs. Emission Channel

The resonance spectrum for shallow band edge trap states detected by band edge excitonic emission, shown in Fig. 4.2c, is composed of three primary Gaussian resonances, two narrow and one broad, which all lead to PL quenching. The broad central resonance ($g_2 \approx 1.96$) is likely due to a single carrier whose wavefunction is somewhat delocalized over the nanorod, thereby experiencing a large distribution of hyperfine and strain fields. Since, in contrast to the mechanism described below, there is no indication of a pair process for resonance 2, we speculate that PL quenching here arises due to an Auger mechanism involving an exciton and a single trapped charge, in analogy to models of blinking in single nanocrystals [11–13]. The two narrower resonances ($g_1 \approx 2.00$ and $g_3 \approx 1.85$) have been attributed to spin-$\frac{1}{2}$ carriers which are localized to the surface of the nanocrystal [25], which limits the range of magnetic environments experienced by the spins and thus environmental spectral broadening. An interesting aspect of these two narrow features is that they have the same resonance area, which describes the probability of undergoing a spin-resonant transition followed by some form of optical activity. In addition, as shown in Fig. 4.2f, the two peaks exhibit exactly the same time dynamics following a microwave pulse. The two peaks must then correspond to electron and hole resonances. Since the two carriers in the pair are correlated by spin-dependent recombination, the resonance of either of these two species leads to the same overall change in "bright-to-dark" exciton population ratio; once perturbed, the system evolves freely to a steady-state condition in exactly the same way at the two magnetic fields (g-factors). This equality of resonant area and time dynamics makes a secure case for these carriers constituting a coupled state.

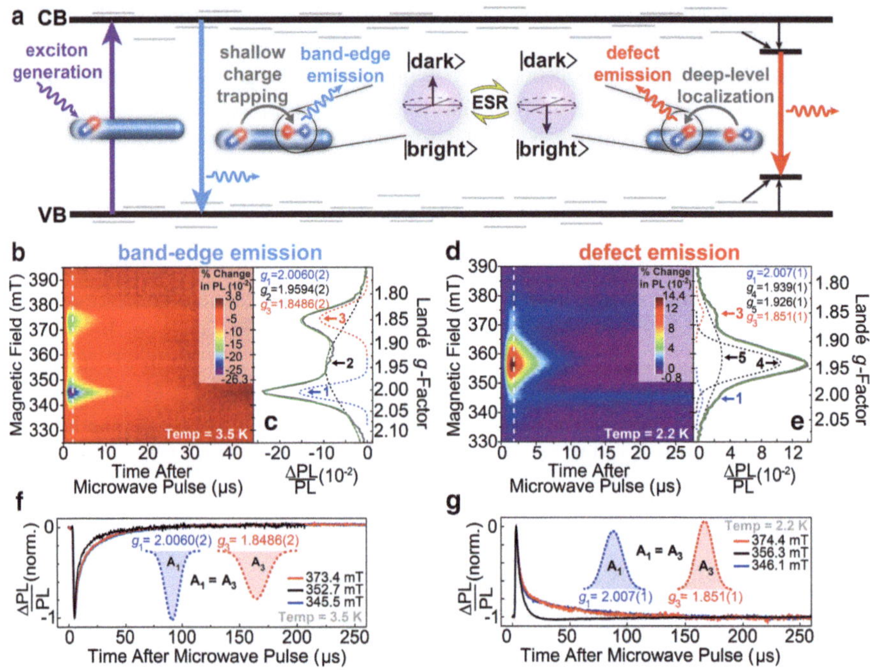

Fig. 4.2 Spectrally resolved ODMR confirms that the correlated trap states feed each emission channel. (**a**) Optical excitation near the band edge populates the lowest exciton state. This exciton either emits, localizes to shallow trap states, or dissipates to the chemical defect. Long carrier trapping lifetimes allow for use of electron spin resonance (ESR) in changing the mutual spin configuration of trapped charge pairs, modulating optically "bright" and "dark" population ratios and thereby affecting the resultant PL intensity from each of the two emission channels. (**b**) Spin resonance mapping and (**c**) resonance spectrum for shallow trap states affecting band gap exciton emission; and (**d, e**) for defect emission. Multiple resonances (i.e. optically active carrier states) are observed through each emission channel. Two resonances ($g_1 \approx 2.00$ and $g_3 \approx 1.85$) are found to be common to both emission channels, indicating that both band edge excitons and emissive chemical defects interact with the same species of shallow band edge trap states. (**f, g**) These two resonances arise due to a coupled carrier pair (i.e. electron and hole), as evidenced by the correlations in resonance peak areas and temporal dynamics for each emission channel. The dynamics of the broad resonance 2 (band edge emission, *black line* in (**f**)) and the superimposed resonances 4 and 5 (defect emission, *black line* in (**g**)) differ from those of the respective pairs (resonances 1 and 3, *blue* and *red lines*, respectively)

It can be shown by analysis of the frequency components observed in coherent Rabi oscillations of the spin species that these signatures arise due to trapped carrier pairs experiencing negligible exchange and magnetic dipole-dipole interactions, but which are still strongly Coulombically bound. Since the external magnetic field lifts the degeneracy, the mutual spin orientation in the pair assumes either singlet or triplet character, but only while these carriers remain trapped and localized. Upon detrapping to band edge exciton states, this singlet-triplet character becomes projected onto the higher spin multiplicity which is well known to characterize the excitonic fine-structure [26, 27]. Therefore, changing the singlet to triplet content of trapped carrier pairs will modify the probability of moving the carrier pair back into one of the three radiative spin-allowed ("bright") or two spin-forbidden ("dark") exciton levels. Ultimately, this conversion of spin multiplicity under resonance changes the overall bright-to-dark state exciton population ratio once detrapping has occurred. Similar dependencies on mutual spin orientation exist for the emissive defect center, although we note that there is no information available on the nature of its excitonic fine structure. The remainder of this work focuses on the spin-resonant dynamics observed in emission from this deep-level trap center.

The magnetic resonance spectrum detected under emission from the defect in Fig. 4.2e exhibits four PL *enhancement* rather than quenching processes. Upon fitting the resonance structure, it is found that the same two coupled states that were observed under detection of the band edge emission channel ($g_1 \approx 2.00$ and $g_3 \approx 1.85$) are also present in resonant modulation of the defect emission, even though these are spectrally entirely distinct species. PL quenching of the band edge emission under resonance correlates directly with PL enhancement of the defect emission. This conclusion is based not only on the equality of g-factors, but also, as before, on the identify of resonance areas and free-evolution dynamics, summarized in Fig. 4.2f, g. Apparently, as carriers within the shallow band edge traps (the exciton "shelving" states) are placed in a mutual spin configuration corresponding to a "dark" state, the probability of charge transfer or relaxation to the emissive defect site is increased [28]. Two important consequences are implied by this result. One is that *both* electrons and holes are transferred to the emissive defect deep-level trap, suggesting that it is actually a defect cluster which can trap both carriers. Secondly, this mechanism serves as a direct observation of a circumvention process for the phonon bottleneck problem [29] of both carrier types. The additional two resonances comprising the primary central feature in the spectrum have previously received limited attention within cw ODMR investigations [20]. In bulk crystalline CdS, it was found that the pronounced lineshape anisotropies of multiple resonances indicate that this emissive defect is indeed a type of donor-acceptor complex [20] which can accommodate both carriers. In contrast, in CdS nanoparticles, this resonance was purported to arise due to strong carrier-pair exchange coupling [21].

4.6 Increased Dipolar Coupling of Shallow Trap States Associated with the Defect

Additional information on the nature of these spin states responsible for the ODMR signal of the deep-level defect can be gained through observing coherent Rabi oscillations arising during application of the microwave field. Specifically, both spin multiplicity [30] as well as exchange [31] and dipolar [32, 33] interactions leave their imprint on the frequency components contained in the oscillation. Transitions between different Zeeman-split m_s levels produce well-defined Rabi frequency components independent of driving field amplitude, whereas components due to interactions depend on the type of interaction and vary with field strength. By sequentially driving the carriers between bright and dark state mutual spin configurations during Rabi flopping, coherent nutation for each of the primary resonance features is demonstrated in Fig. 4.3. Figure 4.3a illustrates the coherent spin propagation scheme employed. Under detection of the defect emission, Rabi flopping is shown in panel (b) for the two individual resonance peaks g_1 (blue) and g_3 (red), and in panel (c) for the overlapping peaks g_4 and g_5 together (black). The Fourier transform of oscillations (raw data inset) is shown with a frequency scale normalized to the free-electron-spin nutation frequency, which is given by the product of the gyromagnetic ratio ($\gamma \approx 28.024\,\mathrm{GHz/T}$) and the strength of the driving microwave field ($B_1 \approx \mathrm{mT}$). The frequency spectrum of such a decaying oscillation exhibits a swept frequency response about the primary frequency component [34]. As reported in Chap. 3, when coherent spin precession for the g_1 and g_3 resonances is read out through the band edge emission channel, there is no signature of either exchange or dipolar interactions and the spin-multiplicity for each carrier is unambiguously $S = \frac{1}{2}$. Surprisingly, when the same trapped carrier pairs are probed coherently, but instead Rabi information is accessed through deep-level chemical defect emission, a clear deviation from the previously observed frequency of γB_1 is noted (see Fig. 3.7 and Sect. 3.6.4). Since spin multiplicity cannot lead to a frequency component lower than γB_1, some change in mutual spin interaction between the two carriers has apparently occurred. Also, exchange interaction produces multiple frequency components [31], not a single low-frequency component. Therefore, it may be presumed that dipolar interaction has been increased for these two trap states [32, 33]. This change in character is likely induced by the proximity of the trap to the emissive defect cluster; the local structural environment of the surface is significantly altered by the S^{2-} and Cd^{2+} vacancies [35], thereby perturbing the more shallow trap states as well. This structural effect is witnessed by the subtle change in linewidth of a few mT for each of these resonances when going from band edge emission detection to defect emission detection (see the Gaussian fits of the lineshapes in Fig. 4.2 f, g; for band edge emission, widths $w_1 = 5.7\,\mathrm{mT}$ and $w_3 = 8.6\,\mathrm{mT}$; for defect emission, $w_1 = 6.7\,\mathrm{mT}$ and $w_3 = 5.8\,\mathrm{mT}$). By monitoring the g_1 and g_3 resonances through defect emission, we therefore probe only that subset of shallow trap states which is both spatially and energetically associated with the deep cluster defect, thereby modifying the linewidths of the resonance slightly.

The Rabi oscillations taken at the central (g_4 and g_5) resonance feature display a strong frequency component at γB_1, but also at both higher and lower frequencies. Interpreting such information is made difficult due to the fact that the resonance structure being probed actually involves a combination of resonant features (the broad $g_5 \approx 1.93$ and the sharper $g_4 \approx 1.94$ species). Even if the frequency components of each resonance differ, they will become inseparably convoluted without higher-order measurement techniques (such as high-field ESR with control of the crystalline axis, shifting each resonance through orientation-specific crystal-field splitting). This convolution will only occur, though, for resonances where

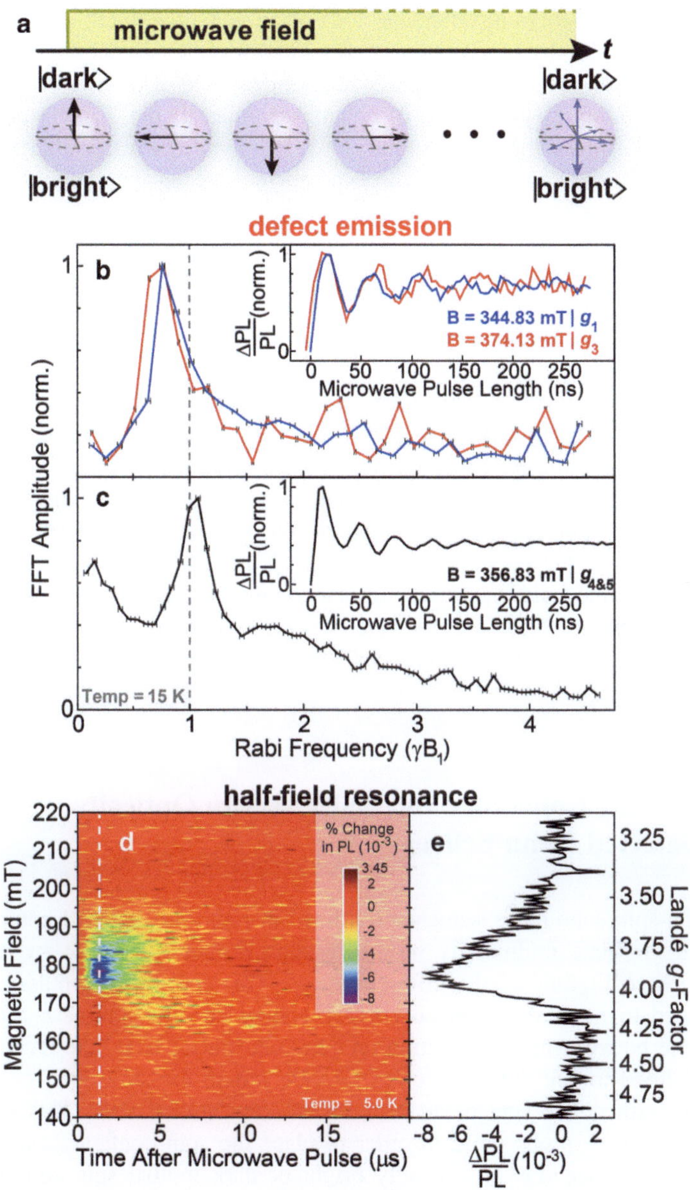

Fig. 4.3 Rabi oscillations and the half-field resonance give evidence for carrier-pair dipolar coupling. (**a**) Coherent Rabi oscillations are driven by a microwave field, which reversibly nutates the spin pair between optically bright and dark mutual spin configurations, enabling read-out of the spin state through the emission intensity. (**b, c**) Fourier transform of the Rabi oscillations of the emissive defect resonances g_1 and g_3 (as identified in Fig. 4.2e); and of the convoluted resonances g_4 and g_5. The insets show the measured Rabi oscillations in the time domain. The frequency components observed in panel (**b**) indicate the occurrence of dipolar coupling, while those in (**c**) are difficult to assign due to the convolution of multiple resonances. (**d**) The half-field resonance confirms the existence of dipolar coupling for at least one of the full-field signals

each feature experiences a sufficiently long coherence time to allow us to measure Rabi oscillations in the experiment (i.e. for $T_2 \geq 10\,\text{ns}$). Consequently, determining whether convolution of the dynamics of resonance species is a factor in the transient spectroscopy requires knowledge of coherence times. If the T_2 time for each of these two resonances differ even slightly, then this should be discernible as a double exponential decay in a Hahn spin echo experiment. Such an experiment is described below, confirming convolution of resonance species.

Since the two satellite features (g_1 and g_3) of the pair process experience dipolar coupling as evidenced by the harmonics observed in the Rabi oscillations, a resonance at approximately half-field ($g_{hf} \approx 4$) is to be expected. Figure 4.3d, e shows that such a resonance is indeed observed. This type of (dipole-forbidden) transition provides evidence of dipolar interactions arising from the $S = 1$ content in at least one set of the full-field transitions. In principle, the features of such a resonance can be used to help establish a rough estimate of spin-pair distances, but information on the corresponding full-field signal is also required (i.e. g-factor, lineshape anisotropy, and spin-orbit coupling tensor), which is lacking here due to the presence of multiple resonances and their convolution. Finally, we note that there is no detectable half-field signal associated with the ODMR gained by monitoring the CdS band edge emission. Strong dipolar coupling of the spin pairs can therefore only arise when these pairs are associated with emission from the deep-level defect, implying that in such species, trapped electron and hole are spatially strongly correlated.

4.7 Probing Coherence and ESEEM with Optically Detected Hahn Echoes

The Hahn spin echo pulse sequence (outlined in Fig. 4.4a) is relied upon in order to reveal the state multiplicity of the resonance about $g = 1.94$ by means of the coherence lifetime of the different states. This conventional pulse sequence is designed to measure the persistence of coherence of a spin as a function of delay time between an initialization and an echo pulse. Here, the technique has been modified to suit the specifics of ODMR, which measures spin permutation symmetry rather than spin polarization as in conventional ESR. This adaptation requires a final $\frac{\pi}{2}$-probe pulse in order to place the spin configuration back into an observable state (i.e. an optically bright or dark mutual spin configuration). The results of this measurement confirm that two long-coherence states are indeed probed at the broad resonance about $g = 1.94$ (i.e. g_4 and g_5), leading to the observed double exponential decay in echo magnitude as a function of interpulse delay time as shown in Fig. 4.4b. Remarkably, the coherence of the longer-lived spin species persists into the microsecond timescale. Such long coherence times are reminiscent of diamond N–V centers. Even in diamond, coupling to nearby defects [36] can cause charge fluctuations [37] and dephasing. Such processes also likely occur here.

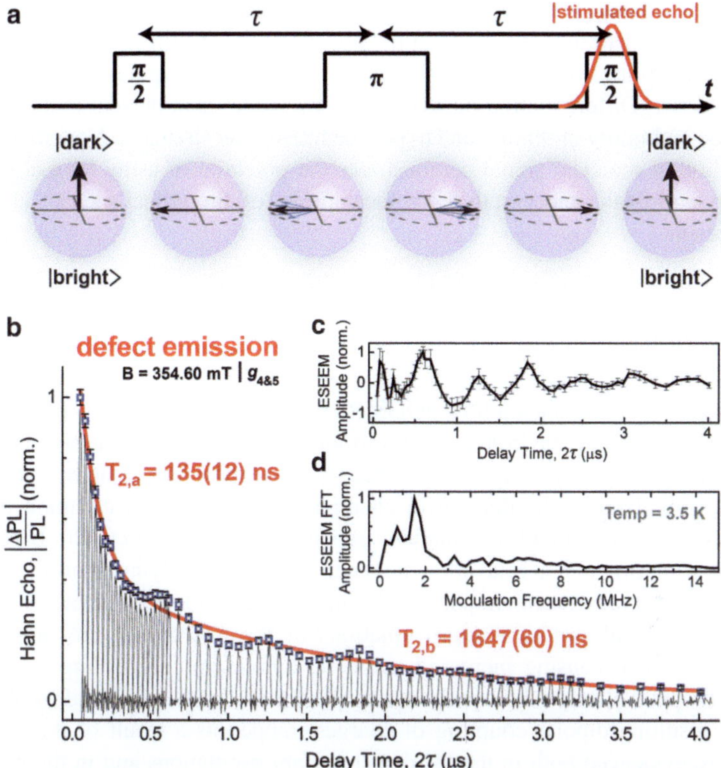

Fig. 4.4 PL-detected Hahn spin echoes reveal slow spin dephasing and an environmental ESEEM signal. (**a**) The conventional Hahn echo pulse sequence is modified for ODMR in order to place the remaining state polarization into an optically observable state. (**b**) The decay of the spin echo recorded in defect emission for the center resonance feature (convolution of peaks g_4 and g_5, as labeled in Fig. 4.2e) is biexponential, suggesting the involvement of two independently resonant carriers under the same resonance condition with distinct dephasing pathways. The very long coherence time of one of these carriers allows probing of the corresponding chemical environment, leading to an electron spin-echo envelop modulation (ESEEM) of the signal. (**c**) ESEEM signal with the biexponential decay removed, and (**d**) corresponding Fourier transform

Nevertheless, this particularly long-lived spin state in semiconductor nanocrystals could find utility in quantum information processing schemes [38]. We note that correlating each coherence time component in the echo signal decay to a respective magnetic resonance could be made possible with selective resonance detection using electron spin echoes observed beyond the shorter coherence lifetime, i.e. by temporally gating out the shortest-lived component.

Additional information on the immediate chemical environment of the spin state can be gained from the long-lived coherence. The pronounced modulation present in the envelope arises due to interactions between the spin of a trapped carrier and its local environment. Such an effect is referred to as electron spin echo envelope modulation, or ESEEM [39, 40]. Figure 4.4c shows the pure contribution of the echo signal due to ESEEM, with the biexponential echo decay removed.

The corresponding Fourier transform of the ESEEM oscillations are given in panel (d) for illustration purposes. In the present case, ESEEM arises due to either hyperfine coupling with Cd nuclear magnetic moments or dipolar coupling with a nearby carrier. Differentiation between these two cases is at present made difficult due to the ambiguity of dipolar and hyperfine interaction strengths. Nonetheless, the modulation of the echo signal demonstrates the potential of using long-coherence states in colloidal nanocrystals as a local probe of the defect's exact chemical environment.

4.8   Conclusion

In this study, we have shown how pODMR can be used as a probe of the chemical nature and electronic environment of various charge trap and emissive chemical defect states. Particular attention has been paid to describing the spin-resonant dynamics involved in the deep-level chemical defect emission common to CdS nanocrystals. It was found that shallow trap states which interact with band edge excitons also provide a relaxation channel to the lower-lying emissive defect state. The observed coherence phenomena in this trap-to-defect relaxation channel indicate that local structural transformations of the nanocrystal are induced by the defect center, causing increased magnetic dipole-dipole coupling among the shallow-trapped electron–hole pairs nearby. The high degree of spatial localization and the resulting dipolar coupling of charges trapped as a result of the emissive defect are evidenced both in the PL-detected Rabi oscillations and in the presence of a pronounced resonance at half-field. Hahn spin echo measurements detected in the emission of the defect itself exposes the extremely long spin coherence lifetime of this center. At $T_2 \approx 1.6\,\mu s$ at 3.5 K, this value is extraordinarily high for colloidal nanocrystals, even compared to magnetically doped particles [39]. In addition, the ESEEM signal opens up future possibilities for gaining insight into the specific chemical and structural information needed in order to engineer this well-known[14] yet incompletely characterized emissive defect out of the CdS nanocrystal synthesis process.

References

1. Semonin, O.E., Luther, J.M., Choi, S., Chen, H.-Y., Gao, J., Nozik, A.J., Beard, M.C.: Peak external photocurrent quantum efficiency exceeding 100 % via MEG in a quantum dot solar cell. Science **334**(6062), 1530–1533 (2011)
2. Dang, C., Lee, J., Breen, C., Steckel, J.S., Coe-Sullivan, S., Nurmikko, A.: Red, green and blue lasing enabled by single-exciton gain in colloidal quantum dot films. Nat. Nanotech. **7**(5), 335–339 (2012)
3. Wood, V., Panzer, M.J., Chen, J., Bradley, M.S., Halpert, J.E., Bawendi, M.G., Bulović, V.: Inkjet-printed quantum dot-polymer composites for full-color AC-driven displays. Adv. Mater. **21**(21), 2151–2155 (2009)

4. Tang, J., Kemp, K.W., Hoogland, S., Jeong, K.S., Liu, H., Levina, L., Furukawa, M., Wang, X., Debnath, R., Cha, D., Chou, K.W., Fischer, A., Amassian, A., Asbury, J.B., Sargent, E.H.: Colloidal-quantum-dot photovoltaics using atomic-ligand passivation. Nat. Mater. 10(10), 765–771 (2011)

5. Zhitomirsky, D., Kramer, I.J., Labelle, A.J., Fischer, A., Debnath, R., Pan, J., Bakr, O.M., Sargent, E.H.: Colloidal quantum dot photovoltaics: the effect of polydispersity. Nano. Lett. 12(2), 1007–1012 (2012)

6. Gómez-Campos, F.M., Califano, M.: Hole surface trapping in CdSe nanocrystals: dynamics, rate fluctuations, and implications for blinking. Nano Lett. 12(9), 4508–4517 (2012)

7. Fischer, S.A., Crotty, A.M., Kilina, S.V., Ivanov, S.A., Tretiak, S.: Passivating ligand and solvent contributions to the electronic properties of semiconductor nanocrystals. Nanoscale 4(3), 904–914 (2012)

8. Voznyy, O.: Mobile surface traps in CdSe nanocrystals with carboxylic acid ligands. J. Phys. Chem. C 115, 15927–15932 (2011)

9. Chestnoy, N., Harris, T., Hull, R., Brus, L.: Luminescence and photophysics of cadmium sulfide semiconductor clusters: the nature of the emitting electronic state. J. Phys. Chem. 90(15), 3393–3399 (1986)

10. Bavykin, D.V., Savinov, E.N., Parmon, V.N.: Surface effects on regularities of electron transfer in CdS and CdS/Cu_xS colloids as studied by photoluminescence quenching. Langmuir 15(14), 4722–4727 (1999)

11. Zhao, J., Nair, G., Fisher, B.R., Bawendi, M.G.: Challenge to the charging model of semiconductor-nanocrystal fluorescence intermittency from off-state quantum yields and multiexciton blinking. Phys. Rev. Lett. 104(15), 157403 (2010)

12. Rosen, S., Schwartz, O., Oron, D.: Transient fluorescence of the off state in blinking CdSe/CdS/ZnS semiconductor nanocrystals is not governed by Auger recombination. Phys. Rev. Lett. 104(15), 157404 (2010)

13. Galland, C., Ghosh, Y., Steinbrück, A., Sykora, M., Hollingsworth, J.A., Klimov, V.I., Htoon, H.: Two types of luminescence blinking revealed by spectroelectrochemistry of single quantum dots. Nature 479(7372), 203–207 (2011)

14. Vuylsteke, A., Sihvonen, Y.: Sulfur cacancy mechanism in pure CdS. Phys. Rev. 113(1), 40–42 (1959)

15. Efros, A.L., Rosen, M.: Random telegraph signal in the photoluminescence intensity of a single quantum dot. Phys. Rev. Lett. 78(6), 1110–1113 (1997)

16. Frantsuzov, P.A., Marcus, R.: Explanation of quantum dot blinking without the long-lived trap hypothesis. Phys. Rev. B 72(15), 155321 (2005)

17. van Driel, A.F., Nikolaev, I.S., Vergeer, P., Lodahl, P., Vanmaekelbergh, D., Vos, W.L.: Statistical analysis of time-resolved emission from ensembles of semiconductor quantum dots: interpretation of exponential decay models. Phys. Rev. B 75(3), 035329 (2007)

18. Jones, M., Lo, S.S., Scholes, G.D.: Quantitative modeling of the role of surface traps in CdSe/CdS/ZnS nanocrystal photoluminescence decay dynamics. Proc. Natl. Acad. Sci. 106(9), 3011–3016 (2009)

19. Jones, M., Lo, S.S., Scholes, G.D.: Signatures of exciton dynamics and carrier rrapping in the rime-resolved photoluminescence of colloidal CdSe nanocrystals. J. Phys. Chem. C 113(43), 18632–18642 (2009)

20. Edgar, A., Pörsch, J.: Optically detected magnetic resonance from a complex donor in CdS. Solid State Commun. 44(5), 741–743 (1982)

21. Lifshitz, E., Litvin, I.D., Porteanu, H., Lipovskii, A.A.: Magneto-optical properties of CdS nanoparticles embedded in phosphate glass. Chem. Phys. Lett. 295(3), 249–256 (1998)

22. Paik, S.-Y., Lee, S.-Y., Baker, W.J., McCamey, D.R., Boehme, C.: T1 and T2 spin relaxation time limitations of phosphorous donor electrons near crystalline silicon to silicon dioxide interface defects. Phys. Rev. B 81(7), 075214 (2010)

23. Kraus, R.M., Lagoudakis, P.G., Rogach, A.L., Talapin, D.V., Weller, H., Lupton, J.M., Feldmann, J.: Room-temperature exciton storage in elongated semiconductor nanocrystals. Phys. Rev. Lett. 98(1), 017401 (2007)

24. McCamey, D.R., Lee, S.-Y., Paik, S.-Y., Lupton, J.M., Boehme, C.: Spin-dependent dynamics of polaron pairs in organic semiconductors. Phys. Rev. B **82**(12), 125206 (2010)
25. Brovelli, S., Galland, C., Viswanatha, R., Klimov, V.I.: Tuning radiative recombination in Cu-doped nanocrystals via electrochemical control of surface trapping. Nano. Lett. **12**(8), 4372–4379 (2012)
26. Efros, A., Rosen, M., Kuno, M., Nirmal, M., Norris, D., Bawendi, M.: Band-edge exciton in quantum dots of semiconductors with a degenerate valence band: Dark and bright exciton states. Phys. Rev. B **54**(7), 4843–4856 (1996)
27. Horodyská, P., Němec, P., Sprinzl, D., Malý, P., Gladilin, V.N., Devreese, J.T.: Exciton spin dynamics in spherical CdS quantum dots. Phys. Rev. B **81**(4), 045301 (2010)
28. Davies, J.: Energy transfer effects in ODMR spectra: a possible source of misinterpretation. J. Phys. C: Solid State Phys. **16**(23), L867–L871 (1983)
29. Schroeter, D., Griffiths, D., Sercel, P.: Defect-assisted relaxation in quantum dots at low temperature. Phys. Rev. B **54**(3), 1486–1489 (1996)
30. Astashkin, A.V., Schweiger, A.: Electron-spin transient nutation: a new approach to simplify the interpretation of ESR spectra. Chem. Phys. Lett. **174**(6), 595–602 (1990)
31. Gliesche, A., Michel, C., Rajevac, V., Lips, K., Baranovskii, S.D., Gebhard, F., Boehme, C.: Effect of exchange coupling on coherently controlled spin-dependent transition rates. Phys. Rev. B **77**(24), 245206 (2008)
32. Limes, M.E., Wang, J., Baker, W.J., Lee, S.Y., Saam, B., Boehme, C.: Numerical study of spin-dependent transition rates within pairs of dipolar and strongly exchange coupled spins with ($s = 1/2$) during magnetic resonant excitation. arXiv:1210.0950 [cond-mat.mtrl-sci]
33. Glenn, R., Limes, M.E., Saam, B., Boehme, C., Raikh, M.E.: Analytical study of spin-dependent transition rates within pairs of dipolar and strongly exchange coupled spins with ($S = 1/2$) during magnetic resonant excitation. arXiv:1210.0948 [cond-mat.mtrl-sci]
34. Glenn, R., Baker, W.J., Boehme, C., Raikh, M.E.: Analytical description of spin-Rabi oscillation controlled electronic transitions rates between weakly coupled pairs of paramagnetic states with $S = 1/2$. arXiv:1207.1754 [cond-mat.mtrl-sci]
35. Wang, Y.R. Duke, C.B.: Cleavage faces of wurtzite CdS and CdSe: Surface relaxation and electronic structure. Phys. Rev. B **37**(11), 6417–6424 (1988)
36. Naydenov, B., Reinhard, F., Lämmle, A., Richter, V., Kalish, R., D'Haenens-Johansson, U.F.S., Newton, M., Jelezko, F., Wrachtrup, J.: Increasing the coherence time of single electron spins in diamond by high temperature annealing. Appl. Phys. Lett. **97**(24), 242511 (2010)
37. Rondin, L., Dantelle, G., Slablab, A., Grosshans, F., Treussart, F., Bergonzo, P., Perruchas, S., Gacoin, T., Chaigneau, M., Chang, H.-C., Jacques, V., Roch, J.-F.: Surface-induced charge state conversion of nitrogen-vacancy defects in nanodiamonds. Phys. Rev. B **82**(11), 115449 (2010)
38. Weber, J.R., Koehl, W.F., Varley, J.B., Janotti, A., Buckley, B.B., Van de Walle, C.G., Awschalom, D.D.: Quantum computing with defects. Proc. Natl. Acad. Sci. **107**(19), 8513–8518 (2010)
39. Ochsenbein, S.T., Gamelin, D.R.: Quantum oscillations in magnetically doped colloidal nanocrystals. Nat. Nanotech. **6**(2), 112–115 (2011)
40. Hoehne, F., Lu, J., Stegner, A., Stutzmann, M., Brandt, M., Rohrmüller, M., Schmidt, W., Gerstmann, U.: Electrically detected Electron-Spin-Echo Envelope Modulation: a highly sensitive technique for resolving complex interface structures. Phys. Rev. Lett. **106**(19), 196101 (2011)

Chapter 5
Summary of Work

5.1 Work in Context

Colloidal nanocrystals are a very interesting material system with inherent flexibility that exists well beyond their size tunable band gap. The choice in boundary conditions on their electronic states is quite large given the wide variety of semiconductor materials, heterostructure configurations, and geometric dimensions available. Thus, these nanocrystals serve as a rich playground for observing single, paired, and multicharge dynamics under a range of parameter configurations. Although ferromagnetic doping effects on these electronic states have certainly been a popular avenue of exploration, paramagnetic effects have remained largely overlooked.

The work presented in this dissertation hopefully serves as an example of how powerful electron spin resonance techniques, with pulsed ODMR in particular, can be when applied to this material system. Access to various charging conditions (energetic traps, chemical defects, etc.), while traditionally difficult to probe directly, were readily available in the studies presented in this work. Time-dependent spin resonance effects were particularly useful here, which enabled subtle resonance features to be distinguished and correlated. Also, the surprisingly long phase-coherence time for some of these states allowed for information on charge-carrier localization to be approached, spin identities to be determined, and probes of carrier-pair and environmental coupling to be considered. Beyond the use of optically active paramagnetic states as local environmental probes, quite novel effects can also be demonstrated, such as spatially remote read-out of spin information and spin-dependence in light-harvesting processes. In all, there remain many exciting opportunities for similar work to be explored in the future.

K.van Schooten, *Optically Active Charge Traps and Chemical Defects in Semiconducting Nanocrystals Probed by Pulsed Optically Detected Magnetic Resonance*, Springer Theses, DOI 10.1007/978-3-319-00590-4_5, © Springer International Publishing Switzerland 2013

5.2   Publications to Date

Spin-dependent exciton quenching and intrinsic spin coherence in CdSe/CdS nanocrystals K. J. van Schooten, J. Huang, W. J. Baker, D. V. Talapin, C. Boehme, J. M. Lupton *Nano Lett.* **13**, 65 (2013)

Robust absolute magnetometry with organic thin-film devices W. J. Baker, K. Ambal, D. P. Waters, R. Baarda, H. Morishita, K. J. van Schooten, D. R. McCamey, J. M. Lupton, C. Boehme *Nat. Commun.* **3**, 898 (2012)

Differentiation between polaron-pair and triplet-exciton polaron spin-dependent mechanisms in organic light-emitting diodes by coherent spin beating W. J. Baker, D. R. McCamey, K. J. van Schooten, J. M. Lupton, C. Boehme *Phys. Rev. B* **84** (16), 165205 (2011)

Hyperfine-field-mediated spin beating in electrostatically bound charge carrier pairs D. R. McCamey, K. J. van Schooten, W. J. Baker, S. -Y. Lee, S. -Y Paik, J. M. Lupton, C. Boehme *Phys. Rev. Lett.* **104** (1), 017601 (2010)

Tuning the singlet-triplet gap in metal-free phosphorescent π-conjugated polymers D. Chaudhuri, H. Wettach, K. J. van Schooten, S. Liu, E. Sigmund, S. Höger, J. M. Lupton *Angew. Chem.* **49** (42), 7714–7717 (2010)

Pulsed electrically detected magnetic resonance in organic semiconductors C. Boehme, D. R. McCamey, K. J. van Schooten, W. J. Baker, S. -Y. Lee, S. -Y Paik, J. M. Lupton *Phys. Status Solidi B* **246** (11–12), 2750–2755 (2009)

Light-harvesting action spectroscopy of single conjugated polymer nanowires M. J. Walter, N. J. Borys, K. J. van Schooten, J. M. Lupton *Nano Lett.* **8** (10), 3330–3335 (2008)

MIX
Papier aus verantwortungsvollen Quellen
Paper from responsible sources
FSC® C105338

If you have any concerns about our products,
you can contact us on
ProductSafety@springernature.com

In case Publisher is established outside the EU,
the EU authorized representative is:
Springer Nature Customer Service Center GmbH
Europaplatz 3, 69115 Heidelberg, Germany

Printed by Libri Plureos GmbH
in Hamburg, Germany